Functionalized Two-Dimensional Black Phosphorus and Polymer Nanocomposites as Flame Retardant

Springer Theses

Recognizing Outstanding Ph.D. Research

Aims and Scope

The series "Springer Theses" brings together a selection of the very best Ph.D. theses from around the world and across the physical sciences. Nominated and endorsed by two recognized specialists, each published volume has been selected for its scientific excellence and the high impact of its contents for the pertinent field of research. For greater accessibility to non-specialists, the published versions include an extended introduction, as well as a foreword by the student's supervisor explaining the special relevance of the work for the field. As a whole, the series will provide a valuable resource both for newcomers to the research fields described, and for other scientists seeking detailed background information on special questions. Finally, it provides an accredited documentation of the valuable contributions made by today's younger generation of scientists.

Theses may be nominated for publication in this series by heads of department at internationally leading universities or institutes and should fulfill all of the following criteria

- They must be written in good English.
- The topic should fall within the confines of Chemistry, Physics, Earth Sciences, Engineering and related interdisciplinary fields such as Materials, Nanoscience, Chemical Engineering, Complex Systems and Biophysics.
- The work reported in the thesis must represent a significant scientific advance.
- If the thesis includes previously published material, permission to reproduce this must be gained from the respective copyright holder (a maximum 30% of the thesis should be a verbatim reproduction from the author's previous publications).
- They must have been examined and passed during the 12 months prior to nomination.
- Each thesis should include a foreword by the supervisor outlining the significance of its content.
- The theses should have a clearly defined structure including an introduction accessible to new PhD students and scientists not expert in the relevant field.

Indexed by zbMATH.

More information about this series at http://www.springer.com/series/8790

Shuilai Qiu

Functionalized Two-Dimensional Black Phosphorus and Polymer Nanocomposites as Flame Retardant

Preparation and Properties

Doctoral Thesis accepted by
the University of Science and Technology of China, Hefei,
China

Springer

Author
Dr. Shuilai Qiu
State Key Laboratory of Fire Science
University of Science
and Technology of China
Hefei, Anhui, China

Supervisors
Prof. Yuan Hu
State Key Laboratory of Fire Science
University of Science and Technology
of China
Hefei, Anhui, China

Prof. Richard K. K. Yuen
City University of Hong Kong
Kowloon, Hong Kong

ISSN 2190-5053 ISSN 2190-5061 (electronic)
Springer Theses
ISBN 978-981-16-3551-9 ISBN 978-981-16-3552-6 (eBook)
https://doi.org/10.1007/978-981-16-3552-6

This Springer imprint is published by the registered company Springer Nature Singapore Pte Ltd.
The registered company address is: 152 Beach Road, #21-01/04 Gateway East, Singapore 189721, Singapore

Supervisor's Foreword

In recent years, the discipline of fire science and safety science and engineering in China has made great progress; especially, the related basic and applied basic research has made a lot of new progress. In the process of the development of the discipline, we have more clearly seen the role and importance of flame-retardant materials applied in fire scenes. Therefore, our research group, based on the State Key Laboratory of Fire Science, University of Science and Technology of China, has been engaged in the basic theoretical research of flame retardant materials and the development of application products for more than 10 years.

In this work, the focus of the research is to use a new two-dimensional nanomaterial—black phosphorus as a high-efficiency flame retardant to overcome shortcomings of the traditional two-dimensional nanomaterials with low flame-retardant efficiency, large flue gas production and poor compatibility with the polymer matrix. At the same time, this thesis provides solutions to the fundamental problems in the practical application of black phosphorus, such as how to prepare the two-dimensional black phosphorus to meet the needs of composite material and improve its compatibility with polymer matrix and dispersion, maintain the stability of the black phosphorus itself in the process of application, which are the key to obtain high-performance nanocomposites.

In the work of this publication, I, as the doctoral supervisor of the author of this paper, was mainly responsible for the formulation of the outline of the book and the supplement and modification of the content of the full text. Dr. Shuilai Qiu was responsible for the compilation of the full text as well as the arrangement, modification and proofreading of words and charts.

This book is thanks to the following funds: the National Key Research and Development Program of China (2017YFC0805901, 2017YFC0805904) and the National Natural Science Foundation of China (21374111, 51675502, 21604081).

In order to ensure the accuracy of the experimental results and data, we have been carefully searched and checked for many times, but because of our limited level, there is a small amount of errors and improper place, please readers to give valuable advice.

Hefei, China
May 2021

Prof. Yuan Hu

Abstract

In recent years, black phosphorus (BP) has attracted extensive attention. Studies have shown that black phosphorus has shown good application prospects in energy storage, catalysis, photoelectronics, biomedicine and other fields. Black phosphorus has a honeycomb-like folded lamellar structure and is thermodynamically stable. As a new two-dimensional (2D) material, black phosphorus can be peeled into nanosheet. Similar to 2D graphene, black phosphorene has the potential to be used as a nano-additive to enhance the mechanical properties, thermal stability and flame retardancy of polymer materials, due to its two-dimensional morphology, inherent high strength and outstanding thermal properties. The addition of low content of 2D nano-filler can significantly improve the mechanical and thermal properties of polymer materials. Compared with the allotrope of black phosphorus, the traditional flame-retardant red phosphorus, BP may show higher flame retardant efficiency due to its special geometric characteristics. In the process of research, we need to solve several fundamental problems. How to prepare the 2D black phosphorene in large scale to meet the requirements of composite preparation, and improve its compatibility and dispersion with the polymer matrix, as well as ensure the air stability of black phosphorene, is the key to obtain high-performance 2D black phosphorene/polymer nanocomposites. Therefore, it is of great importance to develop methods suitable for the preparation of 2D black phosphorene/polymer nanocomposites and to systematically study the mechanical, thermal- and flame-retardant properties of the polymer composites.

In this thesis, we first consider the simple preparation of polymer-based nanocomposites, and considering the air stability of the black phosphorene itself, simultaneously, 2D black phosphorene and its surface-functionalized materials were prepared by several different methods. We prepared high-quality black phosphorus crystal by modified gas phase transfer method. Then, by means of surface coating of polyphosphazene, the air stability of black phosphorene was improved and the dispersion of black phosphorene in epoxy resin and the phase interfacial effect of the nanocomposite system were improved. The black phosphorene functionalized by polyphosphazene can be dispersed uniformly in the polymer matrix and showS obvious physical barrier effect. Secondly, the exfoliation and functionalization of black phosphorene were realized simultaneously by means of electrochemical method. The addition of modified black phosphorene with a low amount obviously endowed the polymer materials with excellent mechanical and thermal properties. Finally, black phosphorus was exfoliated and surface functionalized simultaneously by small molecule-assisted ball milling method. The hydroxylated and aminated black phosphorene were obtained for further functionalization and application. The surface-functionalized black phosphorene can improve the interfacial interaction with the polymer matrix and contribute to the further surface reaction and give it new properties. Through these designs, these functionalized black phosphorene can give the polymer materials better mechanical and flame-retardant properties, thus reducing the fire risk of the polymer composite system. The main research work is as follows.

1. The exploration of BP-polymer nanocomposites is rare due to the BP tends to be oxidized in the atmosphere. We developed a cross-linked polyphosphazene-functionalized BP with abundant -NH_2 groups via a one-pot polycondensation of 4,4′-diaminodiphenyl ether (ODA) and hexachlorocyclotriphosphazene (HCCP) on the surface of BP nanosheets. Whereafter, the resulted polyphosphazene-functionalized black phosphorus (BP-PZN) was incorporated into epoxy resin (EP). Due to the surface functionalization, the 2D black phosphorene in epoxy composites presents a completely delaminated and uniform dispersion state. Compared with the addition of pure black phosphorene, the thermal stability, mechanical and flame-retardant properties of polymer composites with functionalized black phosphorene have been significantly enhanced. Strong interfacial interaction (covalent bonding) and comprehensive flame-retardant effect (instinct flame-retardant effect of black phosphorene and polyphosphazene, physical barrier effect of BP nanosheet) are the key to enhance the combustion performance of polymer composites. Meanwhile, the EP/BP-PZN nanocomposites exhibit air stability after exposure to ambient conditions for four monthS. The air stability of the BP nanosheets in EP matrix is assigned to surface wrapping by polyphosphazene and embedding in the polymer matrix as dual protection.
2. The crucial step toward practical application of BP is the scalable preparation of single- or few layer BP nanosheets. We utilized a facile, green and scalable electrochemical strategy for generating cobaltous phytate functionalized BP nanosheets (BP-EC-Exf) where the BP crystal is used as the cathode and phytic acid is served as modifier and electrolyte simultaneously. Moreover, high-performance polyurethane acrylate/BP-EC-Exf (PUA/BP-EC) nanocomposites are easily prepared by a convenient UV-curable strategy for the first time. By means of electrochemical method, the exfoliation and functionalization of black phosphorene were realized simultaneously. Significantly; conclusion of introducing BP-EC-Exf into PUA matrix resulted in enhancements in mechanical properties of PUA in terms of the tensile strength and tensile fracture strain and the distinct suppression on flame retardant of PUA in terms of decreased peak heat release rate and total heat release, lower intensities of pyrolysis products including toxic CO. Moreover, the PUA/BP-EC nanocomposites present air stability after exposure to ambient conditions for four months. This modified electrochemical method toward the simultaneous exfoliation and

functionalization of BP nanosheets provides an efficient approach for fabricating BP-polymer-based nanocomposites.

3. Natural nacre offers an optimized guideline for assembling 2D nanosheets into high-performance nanocomposites with lightweight, high strength and excellent mechanical properties. Inspired by the "brick-and-mortar"-layered structure of nacre, a multifunctional bioinspired nanocomposites of few layer hydroxyl-functionalized black phosphorus (BP-OH) with nanofibrillar cellulose (NFC) were fabricated via a vacuum-assisted filtration self-assembly procedure. Owing to the interfacial interaction between 2D BP-OH and one-dimensional (1D) NFC, the effective synergistic strengthening effect of the novel nacre-like BP-OHx/NFC composite film has been successfully achieved, resulting in maximum tensile strength (214.0 MPa) and tensile fracture strain (23.8%). Moreover, these nacre-like composite films show high thermal stability and good fire resistance. The nacre-inspired approach in this work demonstrates a promising strategy for the design of the high-performance and flexible BP-based composite films.
4. In order to further improve the phase interaction of 2D black phosphorene in the polymer matrix and utilize the phosphorus–nitrogen synergistic effect to improve the flame-retardant efficiency of the polymer composites. The triazine-based covalent organic framework/BP-NH_2 (BP-NH-TOF) nanohybrid was synthesized via in situ condensation polymerization method using BP-NH_2as a template. The sandwich inorganic–organic hybrid flame retardant was added to the EP to improve the thermal and flame-retardant properties of the composites. Through this design, the nanohybrids not only have high specific surface area of 2D materials, but also show synergistic flame-retardant effect in polymer nanocomposites. Compared with the addition of pure BP-NH_2, the thermal stability, flame retardancy and smoke suppression and toxicity reduction of EP/BP-NH-TOF nanocomposites have been significantly enhanced. The significant reduction of the fire hazard was primarily due to the synergistic action between the catalytic effect of BP-NH_2 and physical barrier effect for the both of BP-NH_2 and TOF nanosheets.
5. The flame-retardant properties of polymer composites were improved by the synergistic effect of 2D black phosphorene and traditional flame retardants. MCA supramolecular/BP-NH_2(BP-NH-MCA) nanohybrid was prepared via in situ self-assembly method using BP-NH_2 nanosheet as template. Due to the presence of amino group, BP-NH_2 nanosheet and MCA supramolecular are combined by hydrogen bond interaction. This kind of inorganic–organic hybrid flame retardant can maintain the high specific surface area of the original 2D morphology of BP and show good flame-retardant and smoke suppression and toxicity reduction effects. The properties of epoxy composites with BP-NH-MCA hybrid nano-flame retardant were enhanced mainly due to the interface interaction of hydrogen bond and non-covalent bond, as well as the inherent flame-retardant effect and physical barrier effect of MCA and BP-NH_2 nanosheets.

Keywords Two-dimensional BP · Polymer nanocomposites · Mechanical properties · Thermal stability · Flame retardancy

List of Publications

A. Journal article

1. **Shuilai Qiu**; Yuan Hu; Richard K. K. Yuen, et al., Air-stable polyphosphazene-functionalized few layer black phosphorene for flame retardancy of epoxy resins. *Small* **2019**, *15*(10), 1805175.
2. **Shuilai Qiu**; Yuan Hu; Richard K. K. Yuen, et al., Electrochemically exfoliated functionalized black phosphorene and its polyurethane acrylate nanocomposites: synthesis and applications. *ACS Appl. Mater. Interfaces* **2019**, *11*(14), 13652–13664.
3. **Shuilai Qiu**; Yuan Hu; Richard K. K. Yuen, et al., Self-assembled supermolecular aggregate supported on boron nitride nanoplatelets for flame retardant and friction application. *Chem. Eng. J.* **2018**, *349*, 223–234.
4. **Shuilai Qiu**; Richard K. K. Yuen, et al., In situ growth of polyphosphazene particles on molybdenum disulfide nanosheets for flame retardant and friction application. *Compos. Pt. A-Appl. Sci. Manuf.* **2018**, *114*, 407–417.
5. **Shuilai Qiu**; Yuan Hu; Richard K. K. Yuen, et al., Melamine-containing polyphosphazene wrapped ammonium polyphosphate: A novel multifunctional organic-inorganic hybrid flame retardant. *J. Hazard. Mater.* **2018**, *344*, 839–848.
6. **Shuilai Qiu**; Yuan Hu; Richard K. K. Yuen, et al., Self-standing cuprous oxide nanoparticles on silica@polyphosphazene nanospheres: 3D nanostructure for enhancing the flame retardancy and toxic effluents elimination of epoxy resins via synergistic catalytic effect. *Chem. Eng. J.* **2017**, *309*, 802–814.
7. **Shuilai Qiu**; Yuan Hu; Richard K. K. Yuen, et al., Flame-retardant - wrappedpolyphosphazene nanotubes: A novel strategy for enhancing the flame retardancy and smoke toxicity suppression of epoxy resins. *J. Hazard. Mater.* **2017**, *325*, 327–339.
8. **Shuilai Qiu**; Yuan Hu; Richard K. K. Yuen, et al., Constructing 3D polyphosphazene nanotube@mesoporous silica@bimetallic phosphide ternary nanostructures via layer-by-layer method: synthesis and applications. *ACS Appl. Mater. Interfaces* **2017**, *9* (27), 23027–23038.
9. **Shuilai Qiu**; Yuan Hu; Richard K. K. Yuen, et al., A 3D nanostructure based on transition-metal phosphide decorated heteroatom-doped mesoporous nanospheres interconnected with graphene: synthesis and applications. *ACS Appl. Mater. Interfaces* **2016**, *8* (47), 32528–32540.
10. **Shuilai Qiu**; Yuan Hu et al., Preparation of UV-curable functionalized phosphazene-containing nanotube/polyurethane acrylate nanocomposite coatings with enhanced thermal and mechanical properties. *RSC Adv.* **2015**, *5* (90), 73775–73782.
11. **Shuilai Qiu**; Yuan Hu et al., Effect of functionalized graphene oxide with organophosphorus oligomer on the thermal and mechanical properties and fire safety of polystyrene. *Ind. Eng. Chem. Res.* **2015**,*54* (13), 3309–3319.

B. Conference article

1. **Shuilai Qiu**, Yuan Hu. EFFECT OF FUNCTIONALIZED GRAPHENE OXIDE WITH ORGANOPHOSPHORUS OLIGOMER ON FLAME RETARDANT PROPERTIES OF POLYSTYRENE. 4th International Symposium on Flame-Retardant Materials & Technologies 2016, poster, Changchun, China.
2. **Shuilai Qiu**, Yuan Hu. CONSTRUCTING 3D POLYPHOSPHAZENE NANOTUBE@MESOPOROUS SILICA@BIMETALLIC PHOSPHIDE TERNARY NANOSTRUCTURES FOR FLAME RETARDANT TPU. 2nd Asia-Oceania Symposium on Fire Safety Materials Science and Engineering 2017, poster, Shenzhen, China.
3. **Shuilai Qiu**, Yuan Hu. SELF-STANDING CUPROUS OXIDE NANOPARTICLES ON SILICA@POLYPHOSPHAZENE NANOSPHERES: 3D NANOSTRUCTURE FOR ENHANCING THE FLAME RETARDANCY AND TOXIC EFFLUENTS ELIMINATION OF EPOXY RESINS VIA SYNERGISTIC CATALYTIC EFFECT. The 29th Annual BCC meeting on Fire Retardancy Stamford, CT; May 20th–23rd, 2018, oral presentation.

C. Patent

1. 胡源, 邱水來, 等. 一種仿貝殼結構纖維素納米纖維/黑磷烯複合膜的製備方法. 專利號 201811240830.6
2. 胡源, 邱水來, 等. 一種聚磷腈改性黑磷烯的製備方法及其應用. 專利號 201811240836.3

Acknowledgements

First of all, I would like to thank my supervisors, Prof. Yuan Hu, from the University of Science and Technology of China, and Prof. Richard K. K. Yuen from the City University of Hong Kong. They gave me selfless help and continued support in scientific research. The inculcation of the two professors gave me the motivation to complete the doctoral study. Otherwise, it would be difficult for me to finish my graduation thesis on time. In addition, I sincerely thank Prof. Lei Song, Prof. Zhou Gui, Prof. Kim Meow Liew and Prof. Siuming Lo for their encouragement and help in my scientific research and learning.

I could not successfully complete the graduation thesis without the support and help of many teachers and students in the laboratory. I would like to express my heartfelt gratitude to associate Prof. Weiyi Xing, associate Prof. Xin Wang, associate Prof. Bibo Wang, associate Prof. Weizhao Hu, Dr. Yongchun Kan, Dr. Chao Ma and Ms. Juping Ding from our laboratory. I would also like to thank Saihua Jiang, Bin Yu, Yongqian Shi, Bihe Yuan, Panyue Wen, Ying Pan, Dong Wang, Hua Ge and Lijin Duan who have graduated, thank them for their help in my study. At the same time, I also want to thank my colleagues in fire chemistry group, Haibo Sheng, Wei Wang, Yao Yuan, Jiajia Liu, Yanbei Hou, Xiaowei Mu, Wenwen Guo, Wei Cai, Yan Zhang, Junling Wang, Xia Zhou, Lingxin He, Fukai Chu, Longxiang Liu, Xin Jin, Xianling Fu, Chenyu Wang, Can Liao, Congxue Yao, Yuling Xiao, Wenxiang Tian, Shicong Ma, Tao Zhang, Yifan Zhou, Bin Zou, Xiyun Ren, Zhoumei Xu, Ziyu Jin, Feng Zhou, Wenhua Chen, Zhixin Zhao, Jingwen Wang, Longfei Han and Yuyu Zhao. It is a great honor to meet you, a group of friends. Thank you for your constant support and company. I wish you all have a bright future.

In particular, I would like to thank the teachers and staff of the state key laboratory of fire science, the engineering science experiment center and the physical and chemical science experiment center of the USTC for their help and support in the experiment and test analysis. Based on this, I was able to successfully complete my own research work. There are so many people to be thankful for that I have no space to list them all. I will remember those who helped and supported me when I was in trouble.

Finally, I would like to offer this literature to my family, my grandparents, my parents, my sister and my lover. In my life journey, they are my strong spiritual support and endless motivation.

The work was financially supported by the National Key Research and Development Program of China (2017YFC0805901, 2017YFC0805904), the National Natural Science Foundation of China (21374111, 51675502, 21604081), and a Grant from the Research Grant Council of the Hong Kong Special Administrative Region, China (contract grant number CityU 11301015).

Contents

Nomenclature

2D	Two-dimensional
AFM	Atomic force microscopy
BP	Black phosphorus
CF	Carbon fiber
CN	Carbonitride
COFs	Coeffcients of friction
CVD	Chemical vapor deposition
DMA	Dynamic mechanical analysis
DMF	Dimethylformamide
DMSO	Dimethyl sulfoxide
DTG	Derivative thermogravimetry
EDS	Energy dispersive spectrometer
EP	Epoxy resin
FETs	Field-effect transistors
FRs	Flame retardants
FTIR	Fourier-transform infrared spectroscopy
GO	Graphene oxide
HRR	Heat release rate
LDH	Layered double hydroxides
MCC	Microscale combustion calorimeter
MMT	Montmorillonite
MoS_2	Molybdenum sulfide
NFC	Nanofibrillar cellulose
NMP	N-methyl-2-pyrrolidone
OER	Oxygen evolution reaction
PANI	Polyaniline
PHRR	Peak heat release rate
PP	Polypropylene
PS	Polystyrene
PUA	Polyurethane acrylate
PVA	Poly(vinyl alcohol)
PVC	Polyvinyl chloride

QDs	Quantum dots
SAED	Selected area electron diffraction
SEM	Scanning electron microscopy
SPR	Smoke production rate
Te	Tellurium
TEM	Transmission electron microscopy
T_g	Glass transition temperature
TGA	Thermogravimetric analysis
TG-IR	Thermogravimetric analysis-infrared spectrometry
THR	Total heat release
Ti	Titanium
TMDs	Transition metal dichalcogenides
TOF	Triazine-based covalent organic framework
TSR	Total smoke release
wt%	Weight percent
XPS	X-ray photoelectron spectroscopy
XRD	X-ray diffraction

List of Figures

Chapter 1
Introduction

1.1 Background

With the successful preparation and development of graphene, people's understanding of two-dimensional (2D) materials is deepening, and it is found that 2D materials have many unique chemical and physical properties compared with bulk materials, during this period which new 2D materials are constantly emerging. As a new type of 2D semiconductor material, black phosphorus (BP) has attracted extensive academic research in recent years. BP not only has many advantages of other 2D materials, such as high tensile strength and good photoelectric properties; it also has more specific physical properties than other 2D materials including graphene and molybdenum disulfide (MoS_2). The band gap energy of BP changes with the number of layers, compared with graphene and transition metal sulfide. The band gap energy increases as the thickness decreases. In addition, as a p type semiconductor, BP has high hole mobility. These excellent physical properties are not available in many 2D materials, and researchers can apply these properties to many fields, such as optoelectronics, electrocatalysis, batteries and even semiconductors and sensors. These properties can only be demonstrated at the nanoscale by exfoliating it into black phosphorene, either physical or chemical methods. In the past two years, it has been reported that black phosphorene can be used as a new type of nano-additive with wide application prospects to prepare polymer nanocomposites, and enhance the thermal and mechanical properties of the polymer materials.

Polymer materials have been widely used in various fields of human production and life due to their excellent comprehensive properties. With the rapid development of society and economy, pure polymer materials cannot meet the application requirements of industries, due to the limitations of their physical properties. The main constituent elements of most polymer materials are carbon and hydrogen, which are easily decomposed and burned when heated. When the polymer burns, it releases a large amount of light, heat and toxic smoke, which is the main cause of fire casualties, and great threat to human life and property security. Therefore, the development of flame retardant polymer nanocomposite system with excellent properties has important scientific significance and application value. The application of 2D nanoparticles such as graphene in flame retardant polymers has been a milestone. Compared with traditional additives, a small amount of 2D nanoparticles can significantly enhance the comprehensive performance of the polymers. For example, due to its unique 2D lamellar structure, graphene plays a flame retardant role in polymer composites due to its good physical barrier effect, which reduces the heat release rate of matrix to some extent during combustion. On the other hand, the introduction of graphene can significantly improve the thermal stability, mechanical and electrical properties of materials. In particular, 2D black phosphorene has a graphene-like lamellar structure and is composed of flame retardant phosphorus element, which shows outstanding mechanical properties and unique catalytic effect. It can be used as a nanofiller to enhance flame retardant polymer materials and has potential applications in the field of polymer nanocomposites.

Considering that the low air stability of black phosphorene, which limits its application in polymers. At present, the research on the preparation of polymer nanocomposites using 2D black phosphorene as nano-additive is still at the exploratory stage, and only a few literatures have reported the synthesis and preparation of black phosphorene/polymer system, as well as the systematic research on mechanical properties and flame retardant properties. However, as with other types of inorganic/polymer nanocomposites, dispersion and interfacial interactions are key factors affecting the properties of the black phosphorene/polymer nanocomposites. Therefore, in order to develop high performance black phosphorene based polymer nanocomposites, in addition to maintaining the air stability of black phosphorene, it is important to be able to achieve large-scale exfoliating to prepare few-layers black phosphorene and improve its interfacial compatibility with the polymer matrix through functional modification. Therefore, it is particularly important to develop methods suitable for the preparation of polymer/2D black phosphorene nanocomposites and systematically study the properties of the resulting composites. These works have practical guiding significance for the application of black phosphorene in the field of polymer composites.

1.2 Review on Two-Dimensional Black Phosphorus

1.2.1 Structure and Basic Properties of BP

Among several allotropes of phosphorus, BP has the most stable crystal structure under normal conditions. It is formed by alternating layers of phosphorus, in which the atoms form a folded honeycomb lattice [1].

(a) Crystal structure

2D BP is a inorganic semi-conductive material with layered stacked honeycomb structure, and the distance between layers is 5.4 Å, as van der Waals force between the layers [2]. Layered BP has two unique crystal axes ZZ and AC. The structure is anisotropic. In the top view of the structure along the z-axis in the Fig. 1.1 [1], parallel to the X-axis and shaped like a sawtooth, the direction containing sawtooth valley and sawtooth peak structure is called ZZ direction. Parallel to the Y-axis and shaped like an armchair, the direction containing the armrest and the seat structure is AC direction. ZZ direction and AC direction are significantly different, and the distance between every two P atoms along ZZ direction is 0.33 nm, the distance between every two P atoms along the AC direction is 0.45 nm. 2D BP is an orthorhombic system with four P atoms in its primary protocell and eight P atoms in its conventional protocell, the volume of the idiomatic primitive cell is twice that of the primary primitive cell, and each P atom has five valence electrons in the 3p orbital, which covalently bonds with the surrounding three atoms to reach saturation, leading to sp3 hybridization. The vertical view of 2D BP along the z-axis is a hexagonal structure with bond angle of 96.3° and 102.1°, there are two P atomic layers and two types of P-P bonds in a single BP layer: one is that the bond between a phosphorus atom and its closest neighbour in the same plane with bond length of 0.2253 nm; the other is the bond between the top and bottom of a single layer of BP with bond length of 0.2287 nm [3].

(b) Physical properties of black phosphorus

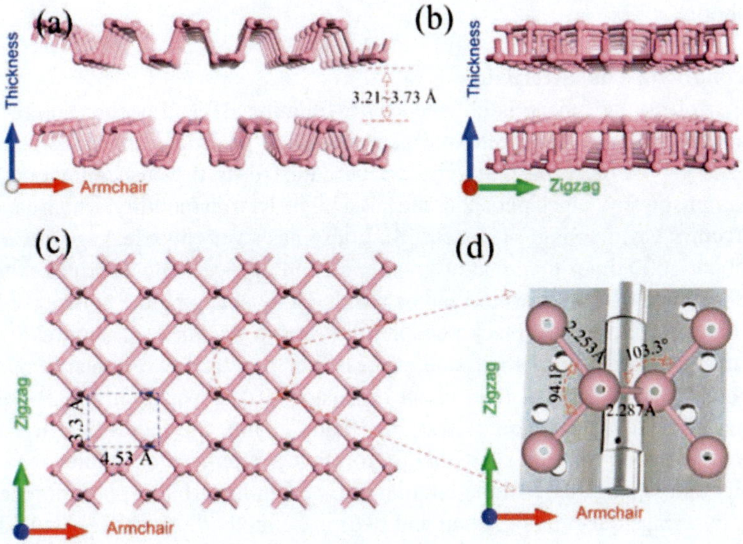

Fig. 1.1 **a, b** Side view from zigzag and armchair orientation, **c** top view of atomic structure of phosphorene, and **d** enlarged local atomic structure of P-P bond configuration [3]

Mechanical properties

The unique stacking structure of 2D BP makes it show excellent mechanical toughness. The young's modulus of single layer BP calculated by the first principle is 44 GPa in the direction of perpendicular to the fold, in the direction parallel to the fold is 92.7 GPa, which is larger than the young's modulus of many conventional materials [4]. Single layer BP can withstand 27% tensile strain along ZZ direction and 30% tensile strain along AC direction, double and multilayer BP can withstand 24% tensile strain along ZZ direction and 32% tensile strain along AC direction. Stretching in different directions is mainly due to the unfolding of the folds rather than the stretching of the bonds. The results show that the breaking strength of few layer BP is greater than 25 GPa and the breaking strain is greater than 8%, which meets the requirements of flexible devices [5].

Thermal properties

The electron conductivity of black phosphorus is relative low, so its thermal transport is mostly covered by phonons. The thermal properties of black phosphorus are also highly anisotropic, and these anisotropes are affected by temperature, strain and size. Studies have shown that the thermal conductivity was size-independent along the direction of the armchair, while the thermal conductivity was size-dependent along the zigzag direction. However, the thermal conductivity decreases significantly in both directions under the action of tensile strain [6]. The thermal conductivity of black phosphorus is equivalent to that of MoS_2 (~1.28 nWK^{-1} nm^{-2}), but lower than that of graphene (~4.1 nWK^{-1} nm^{-2}) [7], due to the folded structure of black phosphorus.

Semiconductor characteristic

Lamellar black phosphene is obtained by exfoliating BP and becomes a 2D semiconductor material with direct band gap composed of ordered phosphorus atoms. The band gap of black phosphorene can be adjusted by the layer number and has obvious anisotropy. Black phosphorene has a high electron mobility, which can reach 10,000 $cm^2/(V\ s)$ for single-layer BP [8]. It also has a very high leakage modulation rate (about 1000 times than that of graphene), and its electronic structure exhibits a direct band gap of 1.5 eV, which can be regulated by changing the number of layers [9]. Zhang et al. obtained black phosphorene with a thickness of about 7.5 nm by mechanical exfoliation method, and successfully prepared two-dimensional black phosphorous single crystal field effect transistor [10]. It was found that when the thickness of black phosphorene was less than 7.5 nm, the field effect transistors prepared had good element switching performance. Koenig et al. studied the carrier mobility and leakage current modulation rate of thin layer black phosphorene transistors by using four-probe method and two-probe method respectively [11]. It was found that the carrier mobility of phosphorene field effect transistor was up to 300 $cm^2/(V\ s)$ at room temperature, and the leakage current modulation rate was over 103. The switching ratio of the transistor is higher than 105 times at low temperature, and the conduction characteristics of electrons and holes are shown. Therefore, the black phosphorus has a broad prospect of research and application in field effect transistor.

Optical properties
There is obvious anisotropy in the optical properties of 2D BP. Through measuring and analyzing the optical absorption spectrum of the few-layer BP, it is found that its saturation and absorption characteristics are better than those of grapheme [12]. And it shows strong linear dichroism along the direction of AC and ZZ. Few-layer BP absorbs polarized light along the AC direction and passes through partial vibration along the ZZ direction, is a natural optical linear polarizer. The photoluminescence of few-layer BP depends on its thickness. Although the smaller the number of layers, the smaller the volume of BP, however, the intensity of photoluminescence of BP increases exponentially with the decrease of the number of BP layers, the photoluminescence intensity of double-layer BP is higher than that of five-layer BP [13].

Chemical stability
Few-layer BP is easy to react with oxygen and water under normal temperature and pressure. With the extension of exposure time in air, the surface of few-layer BP will gradually become rough and a large number of bubbles are generated, such instability limits the application of BP devices [14, 15]. Thus, the degradation mechanism of BP has been deeply studied in order to improve the stability of BP. Atomic force microscopy (AFM) and optical microscopy showed the droplet structure on the surface of BP exposed to environmental conditions [16]. Therefore, the encapsulated layer has been utilized to cover the BP plate and improve its air stability, but water and oxygen may access through the interface, causing the ultimate collapse. Consequently, many scholars are dedicated to finding effective methods for the passivation of BP. The mechanism of BP degradation is that the surface of BP is first oxidized into P_xO_y and then further oxidized to phosphoric acid [17]. The oxygenation rate of BP increases linearly with light intensity and oxygen concentration, and the degradation rate is related to the thickness. The quantum confinement effect becomes more obvious, as the number of BP layers reduces, and the degradation rate of BP is accelerated.

1.2.2 Preparation of Bulk Black Phosphorus

(a) High-pressure method
In the nineteenth century, the scientists argued about BP's existence as a mixture of trace amounts of metallic phosphorus with ordinary phosphorus, which endow it with dark color. Since Percy W. Bridgman prepared the allotrope of P more than one hundred years ago (1914) for the first time, various methods for preparing BP crystals have been explored and developed [18]. Main operating process of this method as follow: The white phosphorus used as reactants and

put it in cylindrical container (about 15 cm long, 1.5 cm inner diameter), the container was put in a special high voltage device for reaction, then increased the unit pressure to 0.6×10^9 Pa, and heat the device at the same time, until the temperature reached 200 °C. Finally, the pressure was increased to the 1.2×10^9 Pa, and the black phosphorus was obtained. This research was mainly concerned with the high-pressure and high-temperature synthesis and properties of materials. Keyes et al. reported the high temperature transformation of white phosphorus to BP under 1.3 GPa, the conversion occurred at 200 °C in a few minutes, at the same time also reduces the volume [19]. This method produced small crystals of less than 100 μm in size by rapid phase transformation at low temperatures. Narita et al. developed new procedure to grow BP crystal under 1 GPa pressure at 550 °C on basis of the phase transformation of red phosphorus, in order to achieve further improvement in BP crystal growth. As expected, the resulting crystals with dimension of $5 \times 5 \times 10$ mm^3 [20]. Red or white phosphorus can also be converted to BP at room temperature, but extremely high pressure must be applied. The synthesis of BP under room-temperature at 10 GPa was reported by Günther and co-works in 1943 [21]. However, the high-pressure equipments for above preparation methods need to be special customized, so these high-pressure methods will be very expensive and are not suitable for industrial fabrication.

(b) Bismuth—flux method

The preparation of BP single crystals via Bismuth-Flux method was developed by Brown and co-works in 1965 [22]. That is, a method to grow black phosphorus crystals from the solution of bismuth. First, 15% of HNO_3 was used to purify the white phosphorus, and then, the purified white phosphorus was dissolved in bismuth solution, subsequently, the solution was heated to 400 °C and maintained 20 h under this temperature, then cooling to room temperature slowly, the acicular or club-shaped black phosphorus crystal can be obtained [23]. Although BP can be easily prepared by this method, it is well known that white phosphorus is easy to be burned in the air, and it is toxic and unsafe to operate in air condition. However, red phosphorus, is relatively stable in air condition, and cannot be dissolved in the liquid bismuth, which can be used as a reactant to prepare BP crystal [24].

(c) Mechanical ball-milling method

In order to achieve simple, efficient, non-toxic and large-scale production of black phosphorus. Mechanical ball-milling method is a new approach to prepare BP crystal, different from the traditional method. In 2007, Park and Sohn reported the amorphous red phosphorus was transformed into orthorhombic BP by high energy mechanical milling at room temperature and ambient pressure, as shown in Fig. 1.2 [25]. The temperature can achieve near 200 °C and also the local pressure can reach about 6 GPa during the process of high-energy mechanical milling, the conditions are suitable for growth of nano- or microcrystalline BP and its composite materials. This method is simple and can achieve grams level of BP at a time, but the resultant BP crystal crystalline

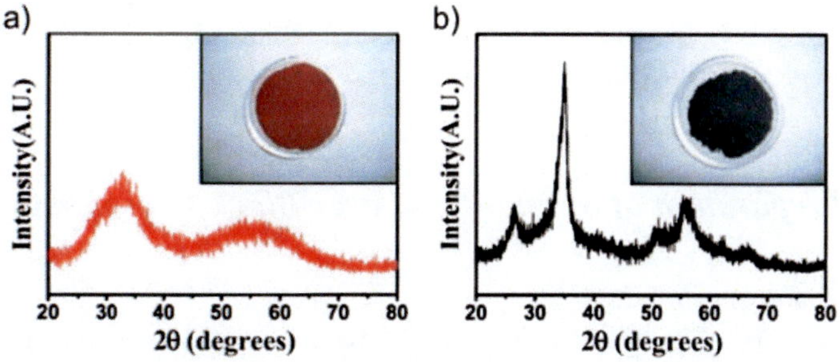

Fig. 1.2 XRD patterns and digital photos of **a** red phosphorus and **b** black phosphorus [25]

has lower crystallinity, the crystal size is below the micron level, which will have a negative impact on its further application.

(d) Chemical gas phase transformation method

Chemical gas phase transformation method is an important method to prepare BP crystals. In 2007, Nilges and his colleagues made remarkable progress in synthesizing BP crystal using Sn, SnI_4, and Au at low temperature [26]. The vapor transport growth method was utilized to obtain BP crystal with red phosphorus as reactant, the SnI_4 and Au–Sn alloy as mineralizer. The specific steps are as follows: First, the mineralizers was mixed with red phosphorus mixed in a certain proportion, then the mixture was sealed in a vacuum quartz tube, BP crystal can be obtained through a series of ascending and cooling procedure, but there are a lot of by-products produced in this process, such as Au_3SnP_7, AuSn and Au_2P_3, etc. [27]. However, the above preparation procedure using the precious Au metal, as we know, the price of Au is high, so the method is not suitable for industrial production. Nilges group in 2014 further improved this procedure based on the original preparation process, in quartz glass ampoule, Sn and SnI_4 were used only without Au, the red phosphorus was converted to BP through high temperature treatment, and the annealing temperature is raised from 600 to 650 °C [28]. This method greatly reduces the cost of preparation, and makes it a common method to prepare BP crystal. The phase transformation strategy of synthesizing BP crystal by utilizing Sn/SnI_4 as mineralizers can increase the yield in large scale compared to other methods, the BP crystals, which can now be used for research purposes, cost about 500 €/g [1]. However, the procedure for above method is highly sensitive to the thermal gradient applied as well as the accurate temperatures used in a quartz ampoule in the preparation process. The Sn/SnI_4 ratio, the ampoule volume and red phosphorus loading are important index of BP synthesis. scale up production of BP crystal should carefully optimized the experimental conditions, since excessive load can lead to breakage of ampoules while insufficient red phosphorus load can only lead to red phosphorus and violet phosphorus achieving. In addition, this approach provides the possibility of growing heterostructure or doping other elements, compared with the high pressure single-crystal growth method.

1.2.3 *Preparation of Single-/few-Layer Black Phosphorus*

(a) Mechanical exfoliation

The layers of two-dimensional (2D) nanomaterials are connected to each other through van der Waals forces, the interlayer forces are weak and can be separated by mechanical forces [29]. This mechanical exfoliation method is commonly used to obtain a thin layer of materials. This method is also utilized to exfoliate graphene and MoS_2 [30, 31]. However, mass production of few-layer BP nano-sheet is still in the primary stage, mechanical peeling is one of the effective approaches to achieve high quality few-layer BP nanoflakes. Firstly, single-/few-layer BP nanoflakes was detached from the bulk BP, and then utilized the tape adhesive/separation methods constantly make it thinner, then transferring the thin phosphorene on the solid substrate (such as Si/SiO_2). Chen and co-workers utilized a scotch tape-based mechanical exfoliation method to exfoliate thin BP films from bulk BP crystal, on the basis of these few-layer BP, they fabricated field-effect transistors and found that charge-carrier mobility depending on thickness [10]. For phosphorene, phosphorene with different thickness has different properties, so it is necessary to develop a preparation method that can control the thickness and size of phosphorene. Ni et al. developed a novel approach based on combination of mechanical cleavage and Ar^+ plasma treatment to achieve stable monolayer phosphorene. The thickness of the few-layer BP (reach five layers) can be clearly identified by AFM and optical contrast spectra [32]. The black phosphorene obtained from plasma thinning method which has few defects, and the number of layers can also be controlled, which is very beneficial for its further application. Mechanical exfoliation method is simple, which can obtain high purity, clean, less defects thin black phosphorene. However, this method has obvious disadvantages such as low production, and time and energy consumption. The exfoliated black phosphorene is extremely unstable and exfoliating operation requires a special environment free of water and oxygen.

(b) Ball milling exfoliation method

Ball milling exfoliation method, is well known solid state mechanical exfoliation method, which has been used to peel bulk BP crystal to prepare black phosphorus nanosheets. However, BP is reactive and unstable, it is easy to transform into red phosphorus without adding catalyst, and the high energy ball milling process is not suitable.

(c) Chemical vapor deposition method

Metal substrates based chemical vapor deposition (CVD) has been widely utilized in the device preparation of TMDs and grapheme [33, 34]. However, at present, there are few reports on synthesis of BP via CVD method. Black phosphorus is highly reactive, which is easy to react with oxygen when exposed to air, as a result, the fragile surface of BP is not beneficial for the growth of black phosphorene. Li and co-workers reported CVD method to fabricate a large-area few-layer BP on a flexible substrate, which can reach to 4 mm [35]. In detail, a layer of thin red phosphorus film was deposited on a polyester (PS) substrate, and its transformation into BP occurs in high-pressure multi-anvil cells. Jiang et al. used red phosphorus as the raw material, carbon nanotubes and titanium (Ti) foil as the substrate, and to synthesize a layer of BP, the resultant BP exhibits high electrocatalytic activity for OER [36]. Chemical vapor deposition method has provided the possibility for the large-scale preparation of black phosphorene with good quality and stable products, and laid the foundation for its wide application.

(d) Electrochemical exfoliation method

Both mechanical exfoliation method and chemical vapor deposition method can successfully synthesize black phosphorene, but both of them have corresponding shortcomings, such as time-consuming, low yield, and so on. Therefore, researchers need to explore and develop a method to produce high-quality black phosphorene in large scale. The previous studies have shown that electrochemical exfoliation can be used to large-scale prepare graphene and molybdenum disulfide with larger lateral sizes [37, 38]. This method is characterized by high efficiency, simple operation, mild reaction conditions, low cost and environmental protection. The basic principle of this method is as follows: In the selected electrolyte, the layered material is used as the anode (or cathode), the ions are driven by the electric field to intercalate into the layered material under the action of the applied electric field, thus the material volume expands and the interlayer van der Waals force decreases, and finally, the layered material is separated into a single layer or a multi-layer. Pumera et al. developed an electrochemical exfoliating approach to peel a bulk BP crystal into few-layer BP flakes with reduced thickness, as shown in Fig. 1.3. Specific steps are as follows: Two-electrode system is designed where Pt foil as the cathode and BP flake as the anode in a 0.5 M H_2SO_4 aqueous solution. The parallel distance between the two electrodes is 2 cm, when the positive DC voltage increases to +3 V, a small amount of tiny particles are slowly released from the BP crystal. After two hours, the solution became darker and tiny particles were observed flashing at the bottom of the electrochemical battery [39]. Zhang et al. reported a one-step, ionic liquid-assisted electrochemical exfoliation and synchronous fluorination process to fabricate fluorinated phosphorene, as a result, the highly selective few-layer fluorinated phosphorene of 3–6 atomic layers can be obtained in large scale [40]. Xiao et al. reported an expandable electrochemical method to exfoliate BP nanoflakes with the BP crystal as the cathode, the as-prepared BP nanoflakes with a thickness of 2–7 nm. It is interesting to note that the electrochemical cathodic exfoliation process completely avoid the oxidation of BP, the high-quality BP nanoflakes can be achieved [41].

(e) Liquid phase exfoliation method

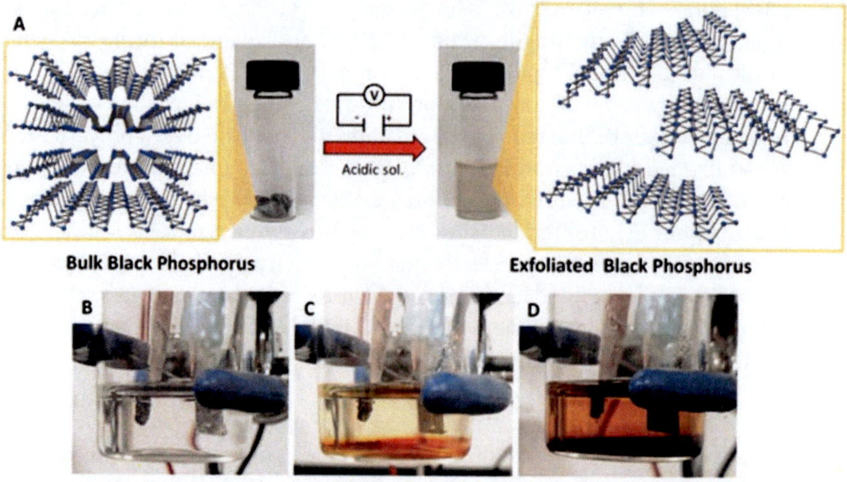

Fig. 1.3 The electrochemical exfoliation procedure. By applying DC voltage, the BP crystal spalling in an H_2SO_4 aqueous solution [39]

Liquid phase exfoliation is also common approach which used to prepare various two-dimensional nanosheets, the bulk material is introduced in the appropriate solvent, when the surface energy of solvent can be matched with the bulk material, the interaction between the solvent molecules and the material can be balance to the energy required for material exfoliation [42, 43]. The bulk materials can be peeled into thin nanosheets by means of ultrasonic oscillation, etc., and the thin layers of the nanosheet can be separated through a suitable centrifugal condition. Recently, this method has been widely used to prepare few-layer BP nanosheets and BP quantum dots [44, 45].

Lewis and O'Brien first reported liquid exfoliation of BP crystal in N-methyl-2-pyrrolidone (NMP) to achieve single or few-layer BP nanoflakes with different lateral dimensions and further studied the exfoliation approach to control size distribution of nanoflakes [46]. Since the first preparation of black phosphorene by liquid-phase exfoliation method, various solvents were selected as the solvents for liquid-phase exfoliating, which can be divided into three categories: Organic solvents, water and ionic liquids. Salehi-Khojin et al. used dimethyl sulfoxide (DMSO) and dimethylformamide (DMF) as liquid-phase exfoliation solvents to prepare highly crystalline few-layer black phosphorene in assistance of ultrasonication to destroy the inter-layer van der Waals interaction of BP layers, where the exfoliated BP nanoflakes are well protected against oxidation in the organic solvents [47]. The presence of organic solvents is bound to have an impact on the application of black phosphorus, Lee and co-workers proposed the exfoliation approach of BP crystals with the introduction of typical ionic liquids (ILs) ([Bmim][Tf_2N] and [Emim][Tf_2N]) as green dispersing media, ionic liquids shown both high exfoliation efficiency and

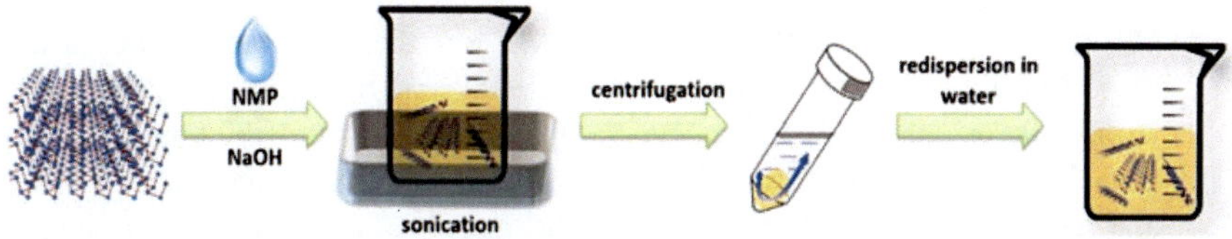

Fig. 1.4 Schematic diagram of the fabrication procedure of basic-NMP-exfoliated phosphorene [49]

higher oxidation resistance in ambient atmosphere of few-layer BP nanoflakes [48]. In order to successfully improve exfoliating efficiency of the BP, Guo and his group developed a liquid method to exfoliate bulk BP in NMP solution with assistance of the basic ion intercalation layer of NaOH, as shown in Fig. 1.4. BP nanosheets with different lateral dimensions were achieved by changing centrifugation speeds [49].

The liquid phase exfoliaiton method also can be used to prepare BP quantum dots. Quantum dots are defined as the nanoparticles with size of several nanometers in three dimensions. Quantum dots have quantum tunneling effect, surface effect, and size effect, etc., are often different from the nanosheets, the preparation and application of quantum dots have always been the focus of research [50]. Sun et al. synthesized black phosphorus quantum dots (BPQDs) by combining bath and probe sonication using liquid exfoliation method, with a thickness of about 1.5 nm and lateral dimension of 2.6 nm, and the ultra-small BPQDs exhibited good photostability as well as an excellent nearinfrared (NIR) photothermal [51].

1.2.4 Surface Modification of Black Phosphorus

Two-dimensional (2D) BP has great application potential in various fields, but the few-layer BP is unstable and easily oxidized and degraded in water or air. Less the layers, and the worse the air stability. The oxidation mechanism of BP has been elucidated that, under the induction of light, oxygen molecules attach on the surface of BP to form superoxide ions (O_2^-). Combination of super oxygen ions and BP will promote the surface of BP layer oxidize to phosphorous oxide (Light provides an advantage to overcome the potential barrier needed for the reaction). In the presence of water, the phosphorus oxides can react quickly with water and degrade into phosphate ions and phosphite ions, the exposure of BP will continue it's oxidation and degradation. Therefore, it is extremely important to inhibit the degradation of BP. The main modification method of BP is the effective surface modification of black phosphorus, including surface layer protection, surface chemical modification, surface etching, etc.

(a) Protective layer

A protective layer is generated on the BP surface by covering the surface to isolate the contact between oxygen, water and BP. Kim et al. reported a double-layer coating which consists of alumina (Al_2O_3) (thickness: 25 nm) and the hydrophobic Teflon-AF fluoropolymer to protect the surface of BP, the results shown that the BP FETs exhibits robust months-long air-stability because of double layer encapsulation with dielectric and fluoropolymer films [16]. Wang et al. proposed a novel and efficient passivation method, the organic perylene-3,4,9,10-tetracarboxylic dianhydride (PTCDA) monolayers were self-assembled on the surface of black phosphorene by using van der Waals epitaxy. When the thickness of PTCDA monolayers reach about 2 nm, PTCDA can isolate black phosphorene from the ambient H_2O and O_2 well. The distinct air-stability of the PTCDA-encapsulated black phosphorene with original unique electronic properties of the raw BP which will promote the opportunity for BP applications in electronic devices [52]. Shao et al. synthesized biodegradable black phosphorus quantum dots/poly (lactic-co-glycolic acid) (BPQDs/PLGA) nanospheres via an emulsion approach by coating BPQDs with PLGA nanoshells. The hydrophobic PLGA can isolate the BPQDs from ambient H_2O and O_2, and control the degradation rate of the BPQDs, thereby improving the photothermal stability. As a result, the photothermal property and absorption curve of PLGA wrapped BPQDs can maintain the air stability above 8 days [53]. Lee et al. reported that the in situ formation of Titanium dioxide (TiO_2) on the surface of BP nanoflakes, the resulting BP@TiO_2 composites exhibited excellent stability in photocatalytic properties. The composites left 92% of catalytic activity after 15 times of repeated irradiation. As a comparison, only 30% of catalytic activity remained for exposed BP after repeated irradiation for 8 times [54].

(b) Surface chemical modification

BP is less stable because of its honeycomb structure, where the phosphorus atom bonds with the other three phosphorus atoms and still has a lone pair of electrons [55], which can be easily taken away by the oxygen molecules, resulting in the oxidation of the outer layer of BP. Preparation of a protective layer can isolate the air and water from BP, but the phosphorus atoms with lone pair electrons are still exist [56]. The lone pair electrons of BP were formed into covalent bond or coordination bond via chemical modification method, blocking its reaction with oxygen can primarily solve the stability problem of BP.

Yang et al. prepared hydroxyl functionalized BP from bulk BP by a mechanochemical ball-milling method, and the LiOH as reactive additive. The introducing of LiOH can terminate the active edge of nanoflakes with hydroxyl groups, which is beneficial to the H_2 evolution activity of visible light for BP nanoflakes [57]. Yang and his group also fabricated BP-C60 nanohybrid by covalent connecting the C60 molecules on the edge of the BP nanoflakes, as shown in Fig. 1.5. The mechanochemical approach by ball-milling C60 powders and bulk BP crystal was utilized to form covalent phosphorus–carbon bonds between C60 molecules and BP nanosheets. As we known, Aromatic

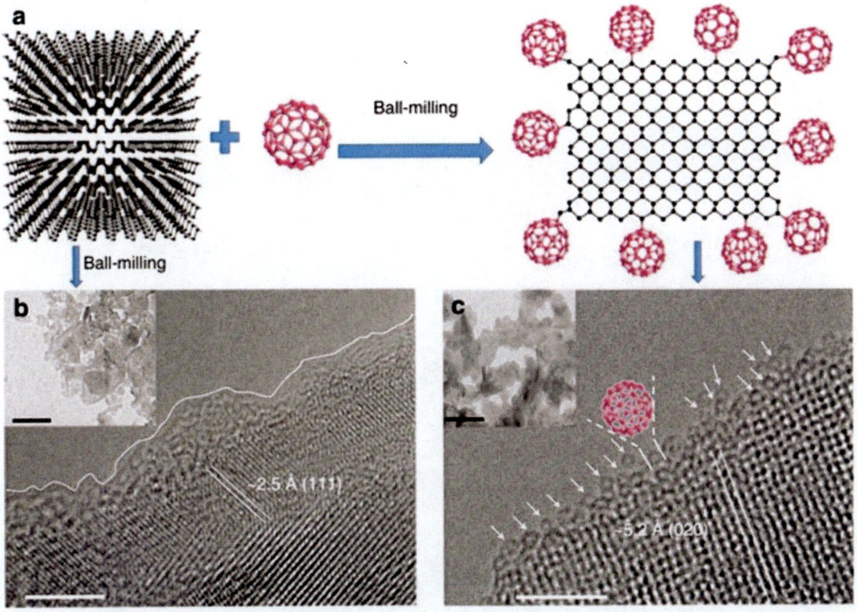

Fig. 1.5 **a** Scheme of the preparation route and structure of BP-C60 hybrid. HRTEM and low-magnification TEM images of **b** the ball-milled BP and **c** the BP-C60 hybrid [58]

and hydrophobic C60 molecules are very stable to oxygen, water, light and other environmental factors. Thus, the edge of BP nanosheet which covalently connected with C60 molecules could efficiently prevent the adsorption of oxygen and water molecules, thereby efficiently enhancing the air stability of BP [58].

For the purpose of enhancing the air stability of BP by preventing oxidization of BP, Chu et al. utilized surface coordination strategy to generate a titanium sulfonate ligand coordinated BP (TiL_4@BP) to improve the stability of BP in humid air and water, as shown in Fig. 1.6. Using the strong electron absorption effect of benzene sulfonate and the vacant orbitals of Ti atom, TiL_4 could coordinate with the lone pair electrons of BP. Ti ligand occupied the lone pair electrons of the BP when a coordination bond was generated, thus they no longer react with oxygen. The untreated BP was unstable, as a comparison, the TiL_4@BP nanohybrid maintained stability in moist air and water (humidity > 90%) for several days [59].

In addition to the modification methods of coordination bond and covalent bond, the surface of BP can be modified by non-covalent bonding method as well. Pumera et al. proposed the noncovalent functionalization method to enhance chemical stabilization of BP by shear exfoliating BP with the redox active antraquinone (AQ), which limited its oxidation in ambient conditions, besides, the additional redox functionality could offer additional charge storage

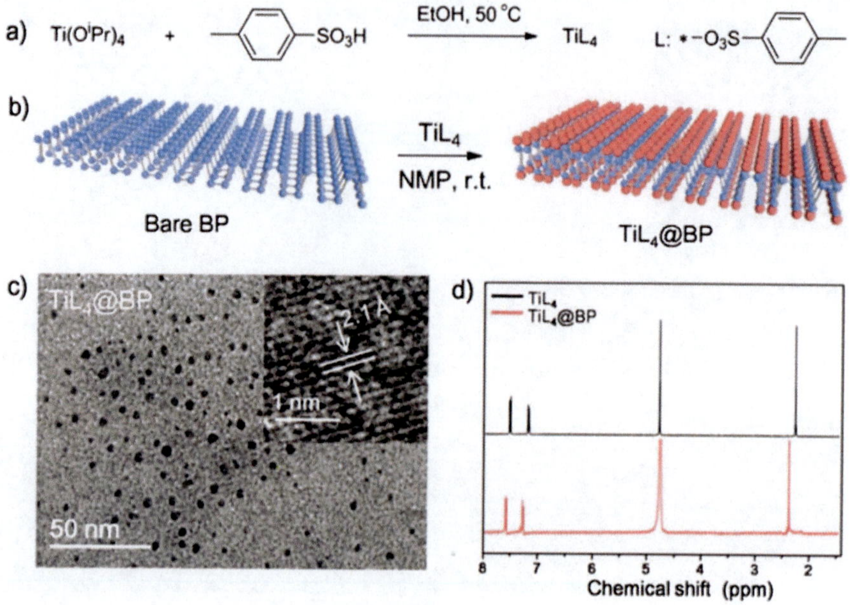

Fig. 1.6 **a** Synthesis and structure of TiL_4; **b** surface coordination of TiL_4 to BP; **c** TEM image of TiL_4@BP; **d** ^1H NMR spectra of TiL_4 and TiL_4@BP [59]

capacity [60]. Hirsch et al. reported that the functionalized BP was achieved on the basis of liquid exfoliation with organic moieties 7,7,8,8-tetracyano-p-quinodimethane (TCNQ). The modification of BP with electron-withdrawing TCNQ resulted in electron transfer from BP to the organic dopant. In addition, the non-covalent interaction of TCNQ with BP was primarily owing to van der Waals forces which resulting in significant air stability of the BP nanosheets against oxidation [61].

(c) Hybridization

One of the main methods to improve the stability of BP is to hybridize with stable materials under air conditions. Pumera et al. reported a facile, large-scale method to synthesize a novel BP/polyaniline (BP/PANI) nanohybrid, as shown in Fig. 1.7. The BP nanosheets play as a template for the formation of PANI. When compared to PANI and pristine BP, the BP/PANI hybrid nanostructure shown outstanding capacitive performance, excellent rate capability, and long-term cycling stability [62]. Cui et al. utilized high energy mechanical milling strategy to fabricate black phosphorus nanoparticle-graphite composites by mechanochemical reaction. During ball-milling process, the phosphorus-carbon bonds were formed, which could promote the stability of the composites during lithium insertion/extraction, keeping outstanding electrical connection between carbon and phosphorus [63]. Peruzzini et al. reported that the nickel/2D black phosphorus (Ni/2DBP) nanohybrid was formed by

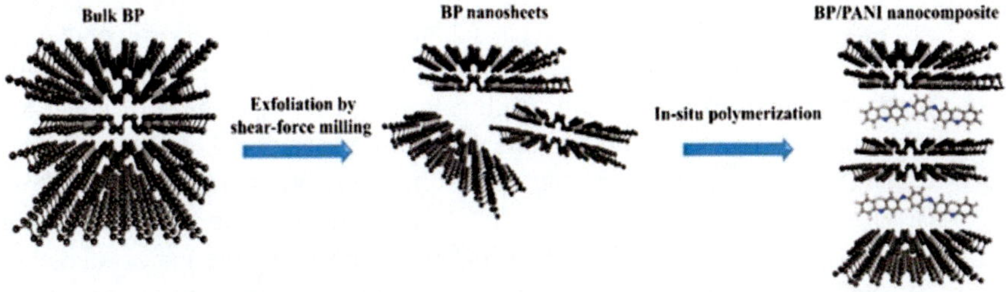

Fig. 1.7 Fabrication process of BP/PANI nanocomposite [62]

dispersing Ni nanoparticles on the surface of exfoliated BP. The as-prepared Ni/2DBP nanohybrid exhibited an enhanced stability compared to pure 2DBP nanosheets when exposure to ambient conditions in the dark. Ni/2DBP was used as a catalyst in the semihydrogenation of phenylacetylene and shown high selectivity and conversion towards styrene. These characters were maintained after recycling tests indicating the high stability of the Ni/2DBP nanohybrid [64]. Hybridizations of BP and different materials provide great advantages for the creation of new functional nanostructures with comprehensive properties. However, as far as current reports are concerned, this method is not as effective as chemically functionalization method in improving the air stability of BP.

(d) Other methods

In addition to the above modifications, doping other elements in black phosphorus can also improve the stability of BP. Liu et al. reported that doping tellurium (Te) in BP can improve the environmental stability and transmission performance of BP devices. Te-doped BP FETs preserve a high mobility of 30% after being exposed to moisture air for three weeks, as comparison, the untreated one shows a lower high mobility of only 2% of the initial value. This work reveal that the element doping is a effective approach to prevent the environmental oxidation of BP, and it will promote the practical application of BP in the field of electronics and optics [65]. The electron mobility and band gap of BP are closely related to the thickness (number of layers) of BP. Surface plasma etching can control the number of BP layers. Kim et al. utilizing fluorine free radicals to etch BP in an Ar atmosphere. The thickness of exfoliated BP nanosheets can be simply controlled by modified plasma treatment, however, the etched BP nanosheets preserve high crystallinity [66].

1.2.5 Applications of Black Phosphorus

(a) Photocatalytic hydrogen productions

Hydrogen is the most ideal renewable energy source. Hydrogen production by photocatalytic water is also a common method [67]. As a catalyst for photocatalytic hydrogen production, it is primary that the band gap is greater than

1.23 eV [68]. Black phosphorus has a band gap of 0.3–2.0 eV depending on the thickness, indicating its potential as a catalyst for photocatalytic hydrogen production. However, due to its high solubility in the presence of water and oxygen, it is extremely limited in its application in the field of photocatalysis. Zhu et al. synthesized a binary nano-mixture (BP/CN) of graphite carbonitride (CN) and 2D BP, and irradiated BP/CN to produce H_2 in water with light, the photocatalytic performance was improved due to the interfacial interaction between CN and BP, thus effective charge transfer occurs [69]. Plasma photocatalytic system is a new type of visible light driven photocatalytic system. Zhu et al. constructed the black phosphorus-sensitized Au/La_2 Ti_2O_7 nanostructure. Compared with the single component catalyst, the catalytic activity of the catalyst was increased by more than 74 and 60 times at wavelength greater than 420 and 780 nm by UV and visible light irradiation. Because of the interfacial interaction between Au/La_2 Ti_2O_7 and BP, which increases the efficiency of electron transfer and enhances the catalytic efficiency [70].

(b) Biological medicines

In the field of biomedicine, biosafety has been widely concerned. The biocompatibility of nanomaterials in biomedical field must be considered. From the angle of biological medicine, there are many advantages to use black phosphorus in the treatment of diseases. First, swollen tumor cells grow speed is more quickly than the normal cells, easy to build into a tumor blood tube in junction structure with defects on the form of a state. The most obvious feature is the large intercellular space in the endothelial cells of the blood vessels, which makes it easy for the nanodrugs through the intercellular space of the endothelial cells of the blood vessels to enter into the tumor tissue and accumulate. At present, black phosphorus nanomaterials mainly include BP nanodots and BP nanosheets. Photothermal therapy (PTT) has the advantages of minimally invasive and high efficiency. However, the clinical application of PTT nanoparticles is limited by biodegradability and long-term toxicity [71]. Experiments have shown that BP has good biocompatibility and applicability for biomedical applications, so BP can be used as a good choice for PTT.

(c) Lithium batteries/solar batteries

As a new favourite of 2D materials, BP has outstanding performance in the field of energy storage. For example, as a battery anode material, the theoretical specific capacity (2596 mAh/g) of BP is about seven times than that of graphite [72]. Two-dimensional materials represented by graphene have fallen into the bottleneck as battery anode materials. These anode materials have lower theoretical specific capacity and shorter life cycle, which greatly limit the development of batteries [73]. Therefore, BP with high theoretical specific capacity and fast charge and discharge capability becomes a potential high performance energy storage material. The negative electrode materials of lithium batteries are more common in graphene and silicon materials, but the theoretical specific capacity of graphene is extremely low (372 mAh/g). Although the theoretical specific capacity of silicon materials is higher (4200 mAh/g), there is a serious volume change (>300%) during battery operation,

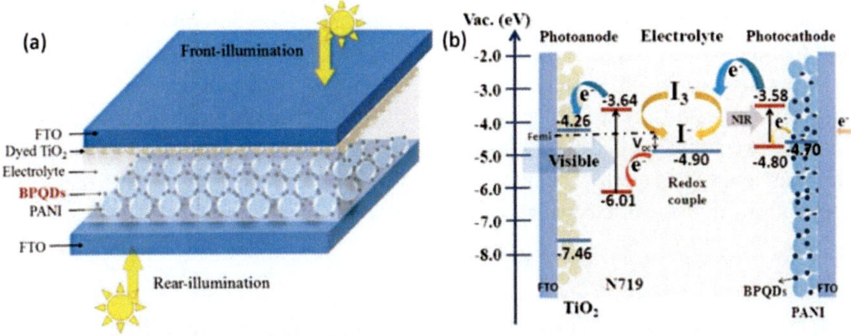

Fig. 1.8 **a** Schematic diagram of double-sides DSSC of photocathode based on BPQDs; **b** diagram for energy level of double-sides DSSCs of photocathode absorption based on BPQDs [76]

which will damage the structure of the battery, resulting in reduced capacity and impact on service life [74]. Chen et al. combined the exfoliated BP nanosheet with highly conductive graphene sheet for application, and proving that BP nanosheet can be used as a high performance flexible LIB electrode similar to paper, exhibiting superior rate performance with high specific capacity of 501 mAh g^{-1} [75].

Yang et al. synthesized BPQD (ultra-small black phosphorus quantum dots) by liquid phase stripping and introduced BPQD into a PANI film to prepare photocathode. The dye-sensitized solar cell (DSSC) with BPQD has a broad-band light absorption and photoelectric efficiency improvement of about 20%, which is mainly owing to the complementary light absorption and fast carrier transport effect of BPQD, as shown in Fig. 1.8 [76].

1.3 Review on Inorganic Layered Compound/polymer Nanocomposites

1.3.1 Introduction

Composite material technology is to combine two or more materials at the molecular level so as to obtain new materials by combining the advantages of several materials. Unlike single-phase nanomaterials, "nanocomposites" consist of two or more gibbs solid phases once one of the dimensions of the composite phase at the nanoscale of 10–100 nm. These solids may be amorphous, semi-crystalline, crystalline, or both. It can be inorganic, organic, or both. The polymer-inorganic nanocomposite materials are divided into inorganic material based composites and polymer based composites. Polymer inorganic nanocomposites play a dominant role in polymer nanocomposites. The combination of inorganic phase and organic phase in the nanometer scale makes the composite material combine many characteristics of inorganic, organic materials and nanoparticles. This is different from the ordinary inorganic filler filling into the polymer. Due to the large surface area of nanoparticles with the quantum effect and surface effect, many physical and chemical properties of the composite material have special changes different from the ordinary filler material. Polymer-based nanocomposites have some special functions, such as luminescence, antibacterial properties, electromagnetic properties, flame retardant properties, etc., thus have many new applications in optical, electrical, mechanical, biological and other fields.

1.3.2 Preparation Method and Structural Characteristics

Conventional preparation methods of inorganic layered compound/polymer nanocomposites include in situ polymerization, solution blending and melt compounding, etc. In-situ polymerization is a method in which the nano-filler 2D nanomaterials are mixed with the monomer or prepolymer of the polymer for polymerization. The prepared nanocomposites generally have good dispersion. When the 2D nanomaterials contain polymeric groups, it can copolymerize with the polymer monomer or prepolymer to form chemical bonds, thus resulting in strong interfacial interaction. In-situ polymerization can occur in the aqueous phase or in the organic phase, and monomers can undergo free radical polymerization and condensation polymerization in the organic phase, which is suitable for the preparation of most polymer-based organic–inorganic nanocomposite systems. Due to the polymer monomer molecules are small and the viscosity is low, the surface modified 2D nanofillers can be uniformly dispersed in the polymer to ensure the uniformity and various physical properties of the composite system. In-situ polymerization is usually used to prepare polymer composites in which that there exists covalent or non-covalent bonds between polymer monomers and 2D nanofillers, such as functionalized graphene/polyurethane nanocomposites [77, 78]. However, due to the presence of 2D nanofillers, the molecular weight of the composites prepared by in-situ polymerization cannot be controlled. Melt compounding methods involves the combining the polymer and 2D nanofiller in the state of polymer melting, generally through the internal mixer, refining machine, screw extruder, etc., is the main processing technology of the current plastic industry. Compared with melt compounding and in-situ polymerization, the dispersion of nanocomposites from melt compounding is generally poor. The decomposition temperatures of some polymers are below the melting point, which limits the type of polymers suitable for this melt compounding process. Compared with other methods, the melt compounding method consumes less energy, and the collision chance of 2D nanofillers increases in the heating process, which makes it easier to agglomerate. Therefore, the surface modification of 2D nanofillers is particular important. Solution blending is to dissolve (disperse) polymers and layered inorganic nanomaterials into an organic solvent, respectively. After fully mixing and stirring for a certain period of time, thus the polymer molecular chain can enter the layer of inorganic substances. In the later stage, the nanosheet is prone to restack in the solvent removal process, in order to solve this problem, covalent or non-covalent modification of 2D nanoadditives before mixing with the polymer solution can effectively prevent the nanosheet from restacking.

1.3.3 Properties

As compared with single phase materials, the interaction between the inorganic nanomaterials and polymers which can produce new effects, to achieve the complementary advantages of these two compounds, and produce more excellent properties. The inorganic layered compound/polymer nanocomposites have been widely used in the fields of reinforced and toughened materials, structural materials, catalytic materials, optical materials, electrical materials and biological materials.

(a) Mechanical property
The introduction of 2D layered nanomaterials into polymers can significantly enhance the mechanical performance of composites. The improvement of mechanical performance of nanocomposites is related to the content, dispersion, ratio of length to diameter, interfacial action and orientation of 2D nanofillers. When the strong interfacial interactions between the nanofiller and polymer matrix are formed, the activity of the polymer chain around the nanofiller is greatly limited, resulting in the formation of the interface hardening phase and the increase of the modulus. As an typical 2D nanofillers, the theoretical young's modulus and fracture strength of graphene can reach 1.1 TPa and 130 GPa, respectively [79], which can be used to effectively enhance the mechanical properties of polymers. Naebe et al. reported that only introducing 0.1 wt% of functionalized graphene into epoxy resin (EP), its flexural strength and storage modulus can be significantly improved. The enhanced mechanical properties of nanocomposites is owing to homogeneous dispersion of graphene and intensive interfacial interaction between functionalized graphene and EP [80]. As another typical 2D nanofiller, the tensile properties of chitosan (CS) matrix can be significantly improved by introducing molybdenum disulfide (MoS_2), the maximum tensile strength of CS matrix can be achieved of 207% by adding only 0.5 wt% of MoS_2. Obviously, even a small quantity of MoS_2 can distinctly enhance the mechanical properties [81].

(b) Thermal property
The thermal stability of polymers is generally poor, and the addition of 2D nanofillers with high thermal stability has a noteworthy positive effect on improving the thermal stability of polymers. The key to enhance the thermal stability of the polymer nancomposites lies in 2D nanofillers' large specific surface area, good dispersion and interaction with the matrices. After addition of 2D nanofillers into polymer matrix to form polymer nanocomposites, 2D nanosheets play as physical barriers and promote to form adiabatic char layer, which can delay the mass loss of polymers, menawhile, inhibit the release of pyrolysis products, and improve the thermal stability of polymer materials.

Song et al. reported at only addition of 0.42 vol% of graphene can remarkably enhance the thermal stability of polypropylene (PP), compared with pure PP, the initial degradation temperature of nanocomposites is enhanced by 26 °C [82]. In addition, the prior works reported that the introduction of montmorillonite (MMT) to PVC matrix decreases the maximum decomposition rate, and increases the initial decomposition temperature. A compact carbon structure is formed on the surface of the composites, which is rich in carbonaceous MMT, can delay the thermal decomposition of the matrix [83].

(c) Gas barrier property

2D nanofillers can significantly reduce the gas permeability of polymer nanocomposites, because the permeation network formed by nanosheet provides a curved channel for gas diffusion. The gas permeability of nanocomposites exists differences while preparing with different 2D nanofillers. Kim et al. reported that the addition of 3 wt% isocyanate modified graphene oxide reduce the nitrogen permeation of thermoplastic polyurethane by about 90% [84]. Cui et al. explored the barrier effect of graphene and its derivatives in the preparation of nanocomposites from different polymer matrix, and the results demonstrated that the graphene layer could generate zigzag paths in the polymer matrix to play the role of gas barrier [85]. The gas barrier properties of 2D nanofillers can inhibit the combustion of polymer materials. As we know, the nanofillers in polymer materials will form a network structure during combustion, which blocking the transmission of pyrolysis gas, thus improving the flame retardant performance of polymer materials. With the improvement of synthesis technology and the development of multiple superposition technology, the perfect barrier performance of 2D nanomaterials has become the research basis of many studies using these nanomaterials as perfect gas-barrier materials.

(d) Flame retardancy

Owing to its high aspect ratio and excellent comprehensive properties, 2D nanomaterials have attracted extensive attention in the application of flame retardant polymer nanocomposites. Compared with conventional halogen-free flame retardants, 2D nanomaterials have obvious advantages. The heat release rate and total heat release of polymer composites can be significantly reduced at a relatively low addition amount (less than 5 wt%). In addition, adding 2D nanofillers can effectively improve the thermal stability and mechanical properties of the polymer nanocomposites system. So far, the typical 2D nanofillers have been reported as effective nano-flame retardants as follow: such as layered metal phosphate [86], layered double hydroxides (LDH) [87], MMT [88], MoS_2 [89], graphene and its derivatives [90]. Bao et al. reported that the flame retardant properties of the functionalized graphene oxide (FGO) based polystyrene (PS) composites were enhanced after incorporating FGO into PS matrix, as a result of the physical barrier effect of FGO and the interface interaction between FGO and the polymers, the peak heat release rate (PHRR) and thermal degradation rate were reduced by 53% and 30%, respectively, and the peak CO release was decreased by 66% [91]. Matusinovic and co-workers used solution mixing techniques to fabricate PS/MoS_2 and poly(methyl methacrylate)/MoS_2 nanocomposites, the PHRR value of PS and poly(methyl methacrylate) composites were reduced by 36% and 24%, respectively, after introducing 10 wt% of MoS_2 [92]. Beyond the physical barrier effect of layered compounds in polymer matrix, other mechanisms for a particular 2D nanomaterial have been studied. For example, the flame retardant mechanism of MoS_2: First, physical barrier effect of MoS_2 nanosheets can slow down the mass loss rate and heat release rate in the combustion process; second, the catalytic action of MoS_2 nanosheets can promote carbonization and inhibit the production of smoke.

(e) Other properties

Different kinds of 2D layered nanomaterials have different intrinsic properties, thus they will exhibit various properties in the polymer composites research system. For example, the electrical and thermal conductivity [93], and electromagnetic shielding properties of graphene based polymer nanocomposites system [94]; tribological behavior and lubrication of molybdenum disulfide based polymer nanocomposites systems [95]; and the application of boron nitride based epoxy composite in the field of electronic packaging materials [96]. Other areas include smart materials and biomaterials.

1.3.4 Recent Progress of Black Phosphorus/polymer Composites

BP aroused great interest of researchers in the application field of energy storage, lithium batteries, supercapacitors, and biomedicine. But up to now, little attention of people has been paid to the preparation, properties and applications of BP based polymer composites. By using keyword "black phosphorus, polymer composites" to retrieve the number of papers published in recent years in the "Web of science" database, it can be found that there are few studies on BP based polymer composites.

Recently, due to the excellent physical properties and unique properties of 2D black phosphorene, it has been reported that 2D black phosphorene can enhance the mechanical properties of polymer composites, which is similar to graphene and other layered materials. Cheng et al. reported that BP nanosheets were used as 2D nanoaddtives to improve the mechanical property of poly(vinyl alcohol) (PVA) composites, as shown in Fig. 1.9. Because of the generation of saturated P-O bonds outside the PVA wrapped BP nanoflakes, the prepared BP-PVA composites exhibit superior air stability. Besides, the friction between PVA moleculars and BP nanoflakes resulting in enhanced modulus, strength, and toughness. BP-PVA nanocomposites show the maximum tensile strength can reach 316.9 ± 12.1 MPa. The reinforcing effects of BP nanosheets in PVA composites is better than those of graphene based PVA composites with the same loading, revealing a promising reinforcement of 2D BP in polymers [97].

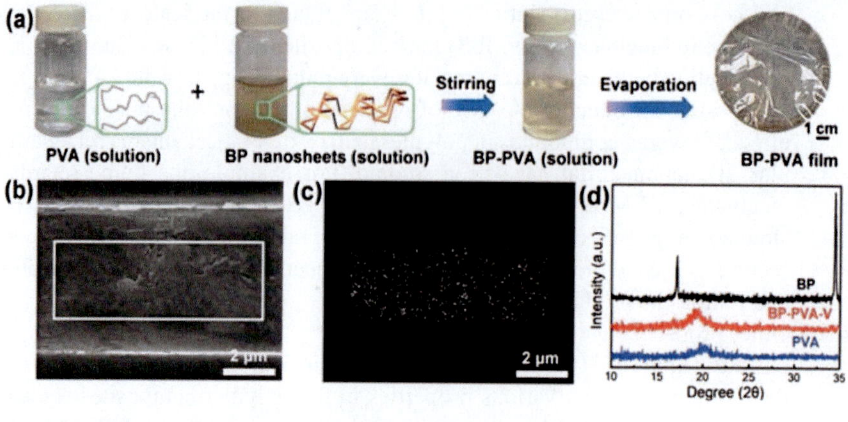

Fig. 1.9 **a** Fabrication procedure of BP-PVA composites; **b** the cross-section morphology; **c** the EDS mapping; **d** XRD patterns of BP-PVA nanocomposites [97]

Black phosphorene has relatively low friction coefficient and high thermal stability. Similar to graphene and MoS_2, black phosphorene can also be used as solid lubricants for the preparation of friction-resistant polymer nanocomposites. Luo et al. fabricated BP nanosheets containing polyetheretherketone (PEEK)/polytetrafuoroethylene (PTFE) and carbon fber (CF)/PTFE composites and investigated their tribological properties. After the introduction of BP nanosheets, the coefficients of friction (COFs) of both the PEEK/PTFE and CF/PTFE composites are dramatically reduced, and the composite shows the minimum COF value of 0.04. The lubrication mechanism of the BP based polymer composites as follow: BP nanosheets can continuously enter the contact area and gradually form a BP film composed of phosphoric acid and phosphorous oxide on the corresponding surface instead of forming a transfer film. The resulted BP containing transfer film promoted the decrease of friction and the reduction of adhesive wear [98].

In addition, in recent years, there are several works have been reported on the preparation of aerogel by using black phosphorene combining polymer. Zhang et al. demonstrated that green injectable composite hydrogels based on BP nanosheet and cellulose have important application value in photothermal therapy (PTT) against cancer [99]. Tang et al. prepared polydopamine functionalized black phosphorus (pBP) nanosheet and then utilized freezing/thawing method to fabricate pBP based PVA composite hydrogels. The resultant pBP/PVA composite hydrogels show good on demand near-infrared (NIR) drug release properties because of high photothermal conversion efficiency of pBP [100].

Red phosphorus has been used as the traditional flame retardants, during burning, they can be thermal decomposed at high temperatures to form phosphoric acid, which promotes the formation of a thermal-resistant carbon coating on the polymers that retards the transfer of oxygen to the combustion zone [101]. However, in order to improve the flame retardant performance, red phosphorus needs to be added in a large proportion, and the compatibility between flame retardant and matrix material is poor [102]. BP is an allotrope of red phosphorus and has higher thermal stability than red phosphorus. Interestingly, black phosphorene exfoliated from BP is a two-dimensional nanomaterial, which has the same properties as graphene, carbon nanotubes and other nano-flame retardants, with high flame retardant efficiency and low addition [103]. Ren et al. introduced 0.2 wt% of black phosphorene (BP) into waterborne polyurethane (WPU) to fabricate BPWPU composites. The heat flow measured by thermal analysis was distinctly reduced by 34.7%, compared with pure WPU, the LOI of BPWPU was reduced by 2.6%, besides, the PHRR was decreased by 10.3% [104]. The flame retardant mechanism of black phosphorene contains two aspects: Catalyze the char formation and capture free radicals, similar to red phosphorus; but also play a physical barrier effect.

1.4 Aims and Objectives

1. From the point of view of obtaining excellent 2D black phosphorene/polymer composites, develop a simple and efficient method for preparing 2D black phosphorene, which can realize large-scale preparation and easy surface functionalization of BP.
2. By modifying the surface of 2D black phosphorene with appropriate methods, dispersion of black phosphorene in polymer materials and the interfacial interaction between two phases in the polymer composite system can be improved while the air stability of black phosphorene is improved.
3. Introducing 2D black phosphorene into polymer matrix through appropriate method, and investigate the physical barrier effect of 2D black phosphorene on the pyrolysis and combustion of polymer composites.
4. By means of electrochemical method, to achieve *simultaneously* exfoliation and surface functionalization of 2D black phosphorene, and characterize the enhancements of thermal stability, mechanical properties and combustion properties of the polymer composite.
5. By means of ball milling method, to achieve *simultaneously* exfoliation and surface functionalization of 2D black phosphorene, characterize the enhancements of the mechanical properties and thermal barrier properties of the polymer composites. Explore the relationship between composition, structure and properties of composite materials.
6. Prepare a variety of 2D black phosphorene based hybrid materials via in-situ synthesis, investigate the synergistic effect of 2D black phosphorene and nitrogen containing flame retardant on reducing fire hazard of polymers, and explore the flame retardant mechanism of hybrid materials during polymer combustion.

1.5 Methodology

Through previous literature research, it can be found that the current research on black phosphorene based polymer nanocomposites is still in the initial stage. In order to realize the industrial application of 2D black phosphorene/polymer composite system, first of all, on the premise of ensuring the air stability of the black phosphorene, it is the key to obtain the large-scale 2D black phosphorene via simple and efficient methods. Secondly, the interfacial compatibility between the unmodified 2D black phosphorene and the polymer materials is poor, which has a great influence on the comprehensive properties of the black phosphorene/polymer composites. In addition, although it has been reported that 2D black phosphorene is similar to graphene and has the potential as nano-flame retardant of polymer materials, however, the single flame retardant mechanism can not meet the needs of practical applications. In order to solve these technical problems, this paper utilizes several typical methods to exfoliate bulk black phosphorus; the surface of 2D black phosphorene is functionalized covalent or non-covalent modification through molecular design; the 2D black phosphorene as a template, several nitrogen-containing organic flame retardants were in situ formed on the surface of black phosphorene to prepare synergistic nano-hybrid flame retardant. Based on the above design principles, the main works of this paper include the preparation of black phosphorus crystal, the exfoliation of 2D black phosphorene, the preparation and performance characterization of polymer composites, and explore the corresponding combustion mechanism.

Preparation of bulk BP: Firstly, bulk black phosphorus crystal was prepared by modified gas phase transformation methods. The crystal has good quality and high crystallinity, which is beneficial to further exfoliation and functionalization.

Preparation of 2D black phosphorene: In this thesis, there are three kinds of exfoliation methods of 2D black phosphorene, namely, direct liquid-phase ultrasonic stripping method, electrochemical exfoliation method and mechanical ball-milling method. The polyphosphazene functionalized black phosphene was achieved by interfacial electrostatic interaction. The cobaltous phytate functionalized BP nanosheets were achieved by simultaneous exfoliation and functionalization via electrochemical method. Hydroxyl and amino modified black phosphorene was prepared by small molecule assisted ball milling method for further functionalization.

Synthesis of triazine based organic framework/black phosphorene nanohybrid through in situ surface growth. Synthesis of MCA supramolecular/aminated black phosphorus nanohybrid by in-situ self-assembly method. The chemical composition of exfoliated and functionalized 2D black phosphorene is investigated with Fourier transform infrared (FTIR) spectra and X-ray photoelectron spectroscopy (XPS). The composition and structure of exfoliated and functionalized 2D black phosphorene is characterized by Raman spectra and X-ray diffraction (XRD) patterns. The micromorphology of exfoliated 2D black phosphorene is evaluated with atomic force microscopy (AFM), scanning electron microscope (SEM) and transmission electron microscopy (TEM).

Preparation of polymer composites: In order to achieve the well dispersion of 2D black phosphorene and its nanohybrids within polymer matrix, preparation methods of polymer nanocomposites including solution blending, thermal curing of epoxy resin, UV-curing of polyurethane acrylates, and a vacuum-assisted filtration self-assembly procedure. The structure, dispersion state and interfacial interaction are studied with SEM, TEM and XRD.

Performance study: To confirm the reinforcement effect of 2D black phosphorene, thermal stability, flame retardancy and mechanical properties are studied. Thermal property is studied with thermogravimetric analysis (TGA) at nitrogen atmosphere. The flame retardant properties are investigated with cone calorimeter and microscale combustion colorimeter (MCC). The mechanical properties are evaluated by dynamic mechanical analysis (DMA) and electronic universal testing instrument. The smoke toxicities are measured by thermogravimetric analysis/infrared spectrometry (TG-IR) test.

Mechanism investigation: The reinforcement mechanism of mechanical properties is related to the mobility of polymer chains, interface adhesion and dispersion, which can be confirmed by the DMA results and the fracture surface of the sample after tensile test. During the decomposition, the pyrolysis products are studied by TG-IR techniques. After the combustion, the composition and structure of residue char are investigated by FTIR, XRD, Raman spectra, XPS and SEM. In summary, on the basis of analysis results of gaseous and condensed products, the flame retardant mechanism can be proposed.

1.6 Thesis Outline

Firstly, Chap. 1 [1] introduces the structure, physical properties, preparation, surface functionalization and application of BP. The literatures concerned on inorganic layered compound/polymer nanocomposites are reviewed. Finally, the progress on recent application of BP based polymer composites is discussed. Chapter 2 presents the surface encapsulation of BP nanosheets. Polyphosphazene functionalized BP and it based polymer composites are prepared, the polyphosphazene is chosen to improve the air stability of few-layer BP and interfacial interaction between 2D BP and epoxy resin (EP) matrix. The thermal and flame retardancy of functionalized BP based EP composites and the related mechanism are proposed. In Chap. 3, bulk BP is exfoliated by electrochemical method in phytic acid solution to achieve simultaneously exfoliating and cobaltous phytate functionalizing BP nanosheets. The mechanical properties of polyurethane acrylate (PUA) composite film reinforcing with functionalized BP nanosheets are characterized. In Chap. 4, inspired by the "brick-and-mortar" layered structure of nacre, a multifunctional bioinspired nanocomposites of few-layer hydroxyl functionalized black phosphorus with nanofibrillar cellulose were fabricated via self-assembly procedure. The thermal and fire resistance, and mechanical performance of bioinspired composites, as well as each mechanism are discussed. According to the study in the previous chapters, Chap. 5 demonstrates the possibility

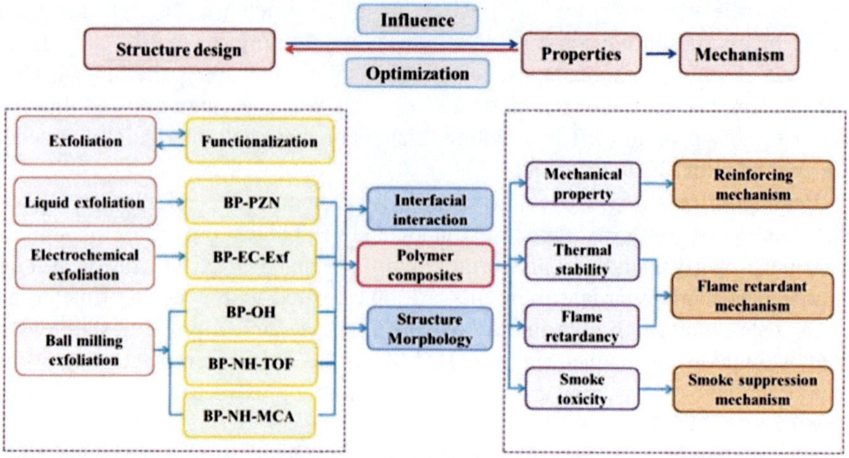

Fig. 1.10 The framework of present dissertation

of amino functionalized BP as a synergist for novel flame retardants triazine based organic framework. By controlling the thermal decomposition in nitrogen atmosphere and comparing the release of pyrolysis products, the physical barrier effect and synergistic effect of triazine based organic framework/BP nanohybrid within EP matrix is indicated. The reinforcement of BP based nanohybrid on mechanical property, thermal stability and flame retardant peroformance of EP material are studied. In Chap. 6, in order to explore the P-N synergistic effect between traditional nitrogenous flame retardant MCA and black phosphorus, MCA supramolecular/BP-NH_2 nanohybrid was prepared by in situ self-assembly method. By studying the pyrolysis and combustion behavior of the composite system, the flame retardant effect and physical barrier effect of this kind of inorganic–organic hybrid flame retardant were revealed. Finally, in Chap. 7, the conclusions, innovations and contributions of this thesis are presented, and recommendations for further research studies are proposed. The framework of present dissertation is depicted in Fig. 1.10.

References

1. Gusmao R, Sofer Z, Pumera M (2017) Black phosphorus rediscovered: From bulk material to monolayers. Angew Chem Int Ed 56:8052–8072
2. Hembram KP, Jung H, Yeo BC, Pai SJ, Lee HJ, Lee K-R, Han SS (2016) A comparative first-principles study of the lithiation, sodiation, and magnesiation of black phosphorus for Li-, Na-, and Mg-ion batteries. Phys Chem Chem Phys 18:21391–21397
3. Kou L, Chen C, Smith SC (2015) Phosphorene: fabrication, properties, and applications. J Phys Chem Lett 6:2794–2805
4. Jiang J-W, Park HS (2014) Mechanical properties of single-layer black phosphorus. J Phys D Appl Phys 47:385304

5. Zhang Z, Li L, Horng J, Wang NZ, Yang F, Yu Y, Zhang Y, Chen G, Watanabe K, Taniguchi T (2017) Strain-modulated bandgap and piezo-resistive effect in black phosphorus field-effect transistors. Nano Lett 17:6097–6103
6. Khandelwal A, Mani K, Karigerasi MH, Lahiri I (2017) Phosphorene–the two-dimensional black phosphorous: properties, synthesis and applications. Mater Sci Eng B 221:17–34
7. Hong Y, Zhang J, Huang X, Zeng XC (2015) Thermal conductivity of a two-dimensional phosphorene sheet: a comparative study with graphene. Nanoscale 7:18716–18724
8. Liu H, Du Y, Deng Y, Peide DY (2015) Semiconducting black phosphorus: synthesis, transport properties and electronic applications. Chem Soc Rev 44:2732–2743
9. Liu H, Zou Y, Tao L, Ma Z, Liu D, Zhou P, Liu H, Wang S (2017) Sandwiched thin-film anode of chemically bonded black phosphorus/graphene hybrid for lithium-ion battery. Small 13:1700758
10. Li L, Yu Y, Ye GJ, Ge Q, Ou X, Wu H, Feng D, Chen XH, Zhang Y (2014) Black phosphorus field-effect transistors. Nat Nanotechnol 9:372
11. Koenig SP, Doganov RA, Schmidt H, Castro Neto A, Oezyilmaz B (2014) Electric field effect in ultrathin black phosphorus. Appl Phys Lett 104:103106
12. Zhang B, Lou F, Zhao R, He J, Li J, Su X, Ning J, Yang K (2015) Exfoliated layers of black phosphorus as saturable absorber for ultrafast solid-state laser. Opt Lett 40:3691–3694
13. Hanlon D, Backes C, Doherty E, Cucinotta CS, Berner NC, Boland C, Lee K, Harvey A, Lynch P, Gholamvand Z (2015) Liquid exfoliation of solvent-stabilized few-layer black phosphorus for applications beyond electronics. Nat Commun 6:8563
14. Wang H, Yang X, Shao W, Chen S, Xie J, Zhang X, Wang J, Xie Y (2015) Ultrathin black phosphorus nanosheets for efficient singlet oxygen generation. J Am Chem Soc 137:11376–11382
15. Kang J, Wood JD, Wells SA, Lee J-H, Liu X, Chen K-S, Hersam MC (2015) Solvent exfoliation of electronic-grade, two-dimensional black phosphorus. ACS Nano 9:3596–3604
16. Kim J-S, Liu Y, Zhu W, Kim S, Wu D, Tao L, Dodabalapur A, Lai K, Akinwande D (2015) Toward air-stable multilayer phosphorene thin-films and transistors. Sci Rep 5:8989
17. Hu Z, Li Q, Lei B, Zhou Q, Xiang D, Lyu Z, Hu F, Wang J, Ren Y, Guo R (2017) Water-catalyzed oxidation of few-layer black phosphorous in a dark environment. Angew Chem 129:9259–9263
18. Bridgman P (1914) Two new modifications of phosphorus. J Am Chem Soc 36:1344–1363
19. Keyes RW (1953) The electrical properties of black phosphorus. Phys Rev 92:580
20. Endo S, Akahama Y, Terada S-I, Narita S-I (1982) Growth of large single crystals of black phosphorus under high pressure. Jpn J Appl Phys 21:L482
21. Günther PL, Geslle P, Rebentisch W (1943) Darstellung und Stabilitätsverhältnisse von schwarzem Phosphor. Z Anorg Allg Chem 250:373–376
22. Brown A, Rundqvist S (1965) Refinement of the crystal structure of black phosphorus. Acta Crystallogr A 19:684–685
23. Maruyama Y, Suzuki S, Kobayashi K, Tanuma S (1981) Synthesis and some properties of black phosphorus single crystals. Physica B + c 105:99–102
24. Baba M, Izumida F, Takeda Y, Morita A (1989) Preparation of black phosphorus single crystals by a completely closed bismuth-flux method and their crystal morphology. Jpn J Appl Phys 28:1019
25. Park CM, Sohn HJ (2007) Black phosphorus and its composite for lithium rechargeable batteries. Adv Mater 19:2465–2468
26. Nilges T, Kersting M, Pfeifer T (2008) A fast low-pressure transport route to large black phosphorus single crystals. J Solid State Chem 181:1707–1711
27. Zhang Y, Zheng Y, Rui K, Hng HH, Hippalgaonkar K, Xu J, Sun W, Zhu J, Yan Q, Huang W (2017) 2D black phosphorus for energy storage and thermoelectric applications. Small 13:1700661
28. Köpf M, Eckstein N, Pfister D, Grotz C, Krüger I, Greiwe M, Hansen T, Kohlmann H, Nilges T (2014) Access and in situ growth of phosphorene-precursor black phosphorus. J Cryst Growth 405:6–10

29. Novoselov K, Jiang D, Schedin F, Booth T, Khotkevich V, Morozov S, Geim A (2005) Two-dimensional atomic crystals. Proc Natl Acad Sci 102:10451–10453
30. Yi M, Shen Z (2015) A review on mechanical exfoliation for the scalable production of graphene. J Mater Chem A 3:11700–11715
31. Li H, Wu J, Yin Z, Zhang H (2014) Preparation and applications of mechanically exfoliated single-layer and multilayer MoS_2 and WSe_2 nanosheets. Acc Chem Res 47:1067–1075
32. Lu W, Nan H, Hong J, Chen Y, Zhu C, Liang Z, Ma X, Ni Z, Jin C, Zhang Z (2014) Plasma-assisted fabrication of monolayer phosphorene and its Raman characterization. Nano Res 7:853–859
33. Tan H, Fan Y, Zhou Y, Chen Q, Xu W, Warner JH (2016) Ultrathin 2D photodetectors utilizing chemical vapor deposition grown WS_2 with graphene electrodes. ACS Nano 10:7866–7873
34. Chen Z, Ren W, Gao L, Liu B, Pei S, Cheng H-M (2011) Three-dimensional flexible and conductive interconnected graphene networks grown by chemical vapour deposition. Nat Mater 10:424
35. Li X, Deng B, Wang X, Chen S, Vaisman M, Karato S-i, Pan G, Lee ML, Cha J, Wang H (2015) Synthesis of thin-film black phosphorus on a flexible substrate. 2D Materials 2:031002
36. Jiang Q, Xu L, Chen N, Zhang H, Dai L, Wang S (2016) Facile synthesis of black phosphorus: an efficient electrocatalyst for the oxygen evolving reaction. Angew Chem Int Ed 55:13849–13853
37. Su C-Y, Lu A-Y, Xu Y, Chen F-R, Khlobystov AN, Li L-J (2011) High-quality thin graphene films from fast electrochemical exfoliation. ACS Nano 5:2332–2339
38. Winchester A, Ghosh S, Feng S, Elias AL, Mallouk T, Terrones M, Talapatra S (2014) Electrochemical characterization of liquid phase exfoliated two-dimensional layers of molybdenum disulfide. ACS Appl Mater Interfaces 6:2125–2130
39. Ambrosi A, Sofer Z, Pumera M (2017) Electrochemical exfoliation of layered black phosphorus into phosphorene. Angew Chem Int Ed 56:10443–10445
40. Tang X, Liang W, Zhao J, Li Z, Qiu M, Fan T, Luo CS, Zhou Y, Li Y, Guo Z (2017) Fluorinated phosphorene: electrochemical synthesis, atomistic fluorination, and enhanced stability. Small 13:1702739
41. Xiao H, Zhao M, Zhang J, Ma X, Zhang J, Hu T, Tang T, Jia J, Wu H (2018) Electrochemical cathode exfoliation of bulky black phosphorus into few-layer phosphorene nanosheets. Electrochem Commun 89:10–13
42. Hernandez Y, Nicolosi V, Lotya M, Blighe FM, Sun Z, De S, McGovern I, Holland B, Byrne M, Gun'Ko YK (2008) High-yield production of graphene by liquid-phase exfoliation of graphite. Nat Nanotechnol 3:563
43. Nicolosi V, Chhowalla M, Kanatzidis MG, Strano MS, Coleman JN (2013) Liquid exfoliation of layered materials. Science 340:1226419
44. Sresht V, Padua AA, Blankschtein D (2015) Liquid-phase exfoliation of phosphorene: design rules from molecular dynamics simulations. ACS Nano 9:8255–8268
45. Xu Y, Wang Z, Guo Z, Huang H, Xiao Q, Zhang H, Yu XF (2016) Solvothermal synthesis and ultrafast photonics of black phosphorus quantum dots. Adv Opt Mater 4:1223–1229
46. Brent JR, Savjani N, Lewis EA, Haigh SJ, Lewis DJ, O'Brien P (2014) Production of few-layer phosphorene by liquid exfoliation of black phosphorus. Chem Commun 50:13338–13341
47. Yasaei P, Kumar B, Foroozan T, Wang C, Asadi M, Tuschel D, Indacochea JE, Klie RF, Salehi-Khojin A (2015) High-quality black phosphorus atomic layers by liquid-phase exfoliation. Adv Mater 27:1887–1892
48. Lee M, Roy AK, Jo S, Choi Y, Chae A, Kim B, Park SY, In I (2017) Exfoliation of black phosphorus in ionic liquids. Nanotechnology 28:125603
49. Guo Z, Zhang H, Lu S, Wang Z, Tang S, Shao J, Sun Z, Xie H, Wang H, Yu XF (2015) From black phosphorus to phosphorene: basic solvent exfoliation, evolution of Raman scattering, and applications to ultrafast photonics. Adv Func Mater 25:6996–7002
50. Shen J, Zhu Y, Yang X, Li C (2012) Graphene quantum dots: emergent nanolights for bioimaging, sensors, catalysis and photovoltaic devices. Chem Commun 48:3686–3699

51. Sun Z, Xie H, Tang S, Yu XF, Guo Z, Shao J, Zhang H, Huang H, Wang H, Chu PK (2015) Ultrasmall black phosphorus quantum dots: synthesis and use as photothermal agents. Angew Chem 127:11688–11692
52. Zhao Y, Zhou Q, Li Q, Yao X, Wang J (2017) Passivation of black phosphorus via self-assembled organic monolayers by van der Waals epitaxy. Adv Mater 29:1603990
53. Shao J, Xie H, Huang H, Li Z, Sun Z, Xu Y, Xiao Q, Yu X-F, Zhao Y, Zhang H (2016) Biodegradable black phosphorus-based nanospheres for in vivo photothermal cancer therapy. Nat Commun 7:12967
54. Lee HU, Lee SC, Won J, Son B-C, Choi S, Kim Y, Park SY, Kim H-S, Lee Y-C, Lee J (2015) Stable semiconductor black phosphorus (BP)@ titanium dioxide (TiO_2) hybrid photocatalysts. Sci Rep 5:8691
55. Hu W, Yang J (2015) Defects in phosphorene. J Phys Chem C 119:20474–20480
56. Lei W, Liu G, Zhang J, Liu M (2017) Black phosphorus nanostructures: recent advances in hybridization, doping and functionalization. Chem Soc Rev 46:3492–3509
57. Zhu X, Zhang T, Sun Z, Chen H, Guan J, Chen X, Ji H, Du P, Yang S (2017) Black phosphorus revisited: a missing metal-free elemental photocatalyst for visible light hydrogen evolution. Adv Mater 29:1605776
58. Zhu X, Zhang T, Jiang D, Duan H, Sun Z, Zhang M, Jin H, Guan R, Liu Y, Chen M (2018) Stabilizing black phosphorus nanosheets via edge-selective bonding of sacrificial C 60 molecules. Nat Commun 9:4177
59. Zhao Y, Wang H, Huang H, Xiao Q, Xu Y, Guo Z, Xie H, Shao J, Sun Z, Han W (2016) Surface coordination of black phosphorus for robust air and water stability. Angew Chem Int Ed 55:5003–5007
60. Gusmão R, Sofer ZK, Pumera M (2018) Functional protection of exfoliated black phosphorus by noncovalent modification with anthraquinone. ACS Nano 12:5666–5673
61. Abellán G, Lloret V, Mundloch U, Marcia M, Neiss C, Görling A, Varela M, Hauke F, Hirsch A (2016) Noncovalent functionalization of black phosphorus. Angew Chem Int Ed 55:14557–14562
62. Sajedi-Moghaddam A, Mayorga-Martinez CC, Sofer ZK, Bouša D, Saievar-Iranizad E, Pumera M (2017) Black phosphorus nanoflakes/polyaniline hybrid material for high-performance pseudocapacitors. J Phys Chem C 121:20532–20538
63. Sun J, Zheng G, Lee H-W, Liu N, Wang H, Yao H, Yang W, Cui Y (2014) Formation of stable phosphorus-carbon bond for enhanced performance in black phosphorus nanoparticle-graphite composite battery anodes. Nano Lett 14:4573–4580
64. Caporali M, Serrano-Ruiz M, Telesio F, Heun S, Nicotra G, Spinella C, Peruzzini M (2017) Decoration of exfoliated black phosphorus with nickel nanoparticles and its application in catalysis. Chem Commun 53:10946–10949
65. Yang B, Wan B, Zhou Q, Wang Y, Hu W, Lv W, Chen Q, Zeng Z, Wen F, Xiang J (2016) Te-doped black phosphorus field-effect transistors. Adv Mater 28:9408–9415
66. Lee G, Lee J-Y, Lee G-H, Kim J (2016) Tuning the thickness of black phosphorus via ion bombardment-free plasma etching for device performance improvement. J Mater Chem C 4:6234–6239
67. Wang X, Maeda K, Thomas A, Takanabe K, Xin G, Carlsson JM, Domen K, Antonietti M (2009) A metal-free polymeric photocatalyst for hydrogen production from water under visible light. Nat Mater 8:76
68. Tijare SN, Joshi MV, Padole PS, Mangrulkar PA, Rayalu SS, Labhsetwar NK (2012) Photocatalytic hydrogen generation through water splitting on nano-crystalline $LaFeO_3$ perovskite. Int J Hydrogen Energy 37:10451–10456
69. Zhu M, Kim S, Mao L, Fujitsuka M, Zhang J, Wang X, Majima T (2017) Metal-free photocatalyst for H_2 evolution in visible to near-infrared region: black phosphorus/graphitic carbon nitride. J Am Chem Soc 139:13234–13242
70. Zhu M, Cai X, Fujitsuka M, Zhang J, Majima T (2017) Au/La_2Ti_2O7 Nanostructures sensitized with black phosphorus for plasmon-enhanced photocatalytic hydrogen production in visible and near-infrared light. Angew Chem Int Ed 56:2064–2068

71. Chen W, Ouyang J, Liu H, Chen M, Zeng K, Sheng J, Liu Z, Han Y, Wang L, Li J (2017) Black phosphorus nanosheet-based drug delivery system for synergistic photodynamic/photothermal/chemotherapy of cancer. Adv Mater 29:1603864
72. Ren X, Lian P, Xie D, Yang Y, Mei Y, Huang X, Wang Z, Yin X (2017) Properties, preparation and application of black phosphorus/phosphorene for energy storage: a review. J Mater Sci 52:10364–10386
73. Sun X-L, Qin X-J, Bo L-M, Wu W (2011) Advances of negative electrode material for lithium ion battery. Nonferrous Metals 2:147–151
74. Xu B (2015) Silicon-based anode materials for lithium-ion batteries
75. Chen L, Zhou G, Liu Z, Ma X, Chen J, Zhang Z, Ma X, Li F, Cheng HM, Ren W (2016) Scalable clean exfoliation of high-quality few-layer black phosphorus for a flexible lithium ion battery. Adv Mater 28:510–517
76. Yang Y, Gao J, Zhang Z, Xiao S, Xie HH, Sun ZB, Wang JH, Zhou CH, Wang YW, Guo XY (2016) Black phosphorus based photocathodes in wideband bifacial dye-sensitized solar cells. Adv Mater 28:8937–8944
77. Kumar M, Chung JS, Kong B-S, Kim EJ, Hur SH (2013) Synthesis of grapheme-polyurethane nanocomposite using highly functionalized graphene oxide as pseudo-crosslinker. Mater Lett 106:319–321
78. Appel A-K, Thomann R, Mülhaupt R (2012) Polyurethane nanocomposites prepared from solvent-free stable dispersions of functionalized graphene nanosheets in polyols. Polymer 53:4931–4939
79. Ji X, Xu Y, Zhang W, Cui L, Liu J (2016) Review of functionalization, structure and properties of graphene/polymer composite fibers. Compos A Appl Sci Manuf 87:29–45
80. Naebe M, Wang J, Amini A, Khayyam H, Hameed N, Li LH, Chen Y, Fox B (2014) Mechanical property and structure of covalent functionalised graphene/epoxy nanocomposites. Sci Rep 4:4375
81. Feng X, Wang X, Xing W, Zhou K, Song L, Hu Y (2014) Liquid-exfoliated MoS_2 by chitosan and enhanced mechanical and thermal properties of chitosan/MoS_2 composites. Compos Sci Technol 93:76–82
82. Song SH, Park KH, Kim BH, Choi YW, Jun GH, Lee DJ, Kong BS, Paik KW, Jeon S (2013) Enhanced thermal conductivity of epoxy–graphene composites by using non-oxidized graphene flakes with non-covalent functionalization. Adv Mater 25:732–737
83. Gong F, Feng M, Zhao C, Zhang S, Yang M (2004) Thermal properties of poly (vinyl chloride)/montmorillonite nanocomposites. Polym Degrad Stabil 84:289–294
84. Kim H, Miura Y, Macosko CW (2010) Graphene/polyurethane nanocomposites for improved gas barrier and electrical conductivity. Chem Mater 22:3441–3450
85. Cui Y, Kundalwal S, Kumar S (2016) Gas barrier performance of graphene/polymer nanocomposites. Carbon 98:313–333
86. Ge H, Tang G, Hu W-Z, Wang B-B, Pan Y, Song L, Hu Y (2015) Aluminum hypophosphite microencapsulated to improve its safety and application to flame retardant polyamide 6. J Hazard Mater 294:186–194
87. Gao Y, Wu J, Wang Q, Wilkie CA, O'Hare D (2014) Flame retardant polymer/layered double hydroxide nanocomposites. J Mater Chem A 2:10996–11016
88. Qin H, Zhang S, Zhao C, Hu G, Yang M (2005) Flame retardant mechanism of polymer/clay nanocomposites based on polypropylene. Polymer 46:8386–8395
89. Wang X, Xing W, Feng X, Song L, Hu Y (2017) MoS_2/polymer nanocomposites: preparation, properties, and applications. Polym Rev 57:440–466
90. Kim M-J, Jeon I-Y, Seo J-M, Dai L, Baek J-B (2014) Graphene phosphonic acid as an efficient flame retardant. ACS Nano 8:2820–2825
91. Bao C, Guo Y, Yuan B, Hu Y, Song L (2012) Functionalized graphene oxide for fire safety applications of polymers: a combination of condensed phase flame retardant strategies. J Mater Chem 22:23057–23063
92. Matusinovic Z, Shukla R, Manias E, Hogshead CG, Wilkie CA (2012) Polystyrene/molybdenum disulfide and poly (methyl methacrylate)/molybdenum disulfide nanocomposites with enhanced thermal stability. Polym Degrad Stabil 97:2481–2486

93. Mamunya YP, Davydenko V, Pissis P, Lebedev E (2002) Electrical and thermal conductivity of polymers filled with metal powders. Eur Polymer J 38:1887–1897
94. Yan DX, Pang H, Li B, Vajtai R, Xu L, Ren PG, Wang JH, Li ZM (2015) Structured reduced graphene oxide/polymer composites for ultra-efficient electromagnetic interference shielding. Adv Func Mater 25:559–566
95. Naffakh M, Díez-Pascual AM, Remškar M, Marco C (2012) New inorganic nanotube polymer nanocomposites: improved thermal, mechanical and tribological properties in isotactic polypropylene incorporating INT-MoS_2. J Mater Chem 22:17002–17010
96. Yu C, Zhang J, Li Z, Tian W, Wang L, Luo J, Li Q, Fan X, Yao Y (2017) Enhanced through—plane thermal conductivity of boron nitride/epoxy composites. Compos A Appl Sci Manuf 98:25–31
97. Ni H, Liu X, Cheng Q (2018) A new strategy for air-stable black phosphorus reinforced PVA nanocomposites. J Mater Chem A 6:7142–7147
98. Lv Y, Wang W, Xie G, Luo J (2018) Self-lubricating PTFE-based composites with black phosphorus nanosheets. Tribol Lett 66:61
99. Xing C, Chen S, Qiu M, Liang X, Liu Q, Zou Q, Li Z, Xie Z, Wang D, Dong B (2018) Black phosphorus hydrogels: conceptually novel black phosphorus/cellulose hydrogels as promising photothermal agents for effective cancer therapy (Adv. Healthcare Mater. 7/2018). Adv Healthcare Mater 7:1870030
100. Yang G, Wan X, Gu Z, Zeng X, Tang J (2018) Near infrared photothermal-responsive poly (vinyl alcohol)/black phosphorus composite hydrogels with excellent on-demand drug release capacity. J Mater Chem B 6:1622–1632
101. Liu Y, Wang Q (2006) Melamine cyanurate-microencapsulated red phosphorus flame retardant unreinforced and glass fiber reinforced polyamide 66. Polym Degrad Stabil 91:3103–3109
102. Lu S-Y, Hamerton I (2002) Recent developments in the chemistry of halogen-free flame retardant polymers. Prog Polym Sci 27:1661–1712
103. Dittrich B, Wartig K-A, Hofmann D, Mülhaupt R, Schartel B (2013) Flame retardancy through carbon nanomaterials: carbon black, multiwall nanotubes, expanded graphite, multi-layer graphene and graphene in polypropylene. Polym Degrad Stabil 98:1495–1505
104. Ren X, Mei Y, Lian P, Xie D, Yang Y, Wang Y, Wang Z (2018) A novel application of phosphorene as a flame retardant. Polymers 10:227

Chapter 2
Air Stable Polyphosphazene Functionalized Few-Layer Black Phosphorene for Flame Retardancy of Epoxy Resins

2.1 Introduction

As a new candidate of the two-dimensional (2D) materials, black phosphorus (BP) has raised concern as a physical analogue of graphite. The atom structure of BP consisting of P atoms covalently bonded to three neighbors with an orthorhombic puckered layer [1]. Since the first exfoliation in 2014, lamellar BP has displayed attractive potential applications owing to various of intriguing properties, for instance, high charge mobility (as high as 1000 cm^2/V s), thickness-tunable bandgap and anisotropic transport [2, 3]. The single- or few-layer BP nanosheets, identified as "phosphorene", rendering its potential wide applications in transistors, storage, energy conversion, and biomedicine [4, 5]. The BP is superior to traditional 2D nanomaterials including graphene and transition metal dichalcogenides. However, BP is exceedingly inclined to oxidation and degradation by ambient species (H_2O/O_2) in the air. The oxidative products, phosphorus oxides (PO_x) may decompose into phosphoric acid after oxidation, which seriously deteriorates its physical and chemical properties [6, 7].

Similar to the graphene from graphite, the key step in the application of BP is the exfoliation of mono- or few-layer phosphorene. Multilayered BP crystal was prepared via high-pressure or high-temperature conversion of red or white phosphorus [8, 9]. Currently, single- to few-layer BP was mainly obtained through top-down strategy by ways of exfoliation procedures, including micromechanical cleavage, ball milling, liquid exfoliation in organic solvents, and plasma thinning treatment. In order to get high quality few-layer BP with no structural damage and a large lateral size, the as-required BP nanosheets exfoliated from bulk BP by micromechanical method achieve the requirement but lacks scalability. The ball milling method produces few-layer BP with edge defects of small lateral size, and the crystallinity become weak. Normally, liquid exfoliation as a optimize choice for large scale exfoliation to acquire high quality single- or few-layer BP with moderate lateral size and few structural damage could strengthen its application. For example, the exfoliation of BP crystal in organic solvent including *N*-methyl-2-pyrrolidone (NMP), this liquid exfoliation approach achieves uniform dispersion in the NMP medium and the scalable exfoliation of BP nanosheets [10, 11]. Therefore, liquid exfoliation methods enable large-scale exfoliation and acquire high quality few-layer BP.

For future applications, numerous research achievements have been acquired for purpose of developing new strategy to obtain air-stable BP [12]. Actually, several efficient approaches have been carried out, including passivation by means of encapsulation with inorganic or organic materials, such as aluminum oxide layer, boron nitride, grapheme [13] and crosslinking polymers [14]. However, these strategies have a positive effect to achieve air stable BP nanosheets, but there exist several defects in terms of high costs, scalable production and complex preparation process. In addition, the encapsulation via the simple sandwiching of BP may have no synergistic strengthen effect for the material properties. Hence, fabricating composites/hybrids of BP and other materials by means of covalent or noncovalent methods could be an optimized way for achieving air-stable BP with improved properties.

Polyphosphazene is a kind of representative organic–inorganic hybrid materials with alternating arrangement of phosphorus (P) and nitrogen (N) in the main chain [15]. The unique structure endow it excellent biocompatibility, water resistance, solvent resistance and chemical modification [16]. Currently, polyphosphazene has been used as flame retardant with superior thermal stability and high limiting oxygen index (LOI) [17]. There are two types of polyphosphazenes, including linear polyphosphazenes and crosslinked polyphosphazenes. In view of crosslinked polyphosphazenes, which have attracted great attention and utilized to fabricate nanoscale polymers by condensation polymerization, for example, polyphosphazene-based nano- or microspheres, nanofibers and nanotubes [18]. In terms of the structural flexibility of -P = N- units and existence of hydroxyl or amino active groups, crosslinked polyphosphazenes are better choices for surface functionalization, microcapsule material and reinforcing additives [19]. To illustrate, polyphosphazene encapsulated ammonium polyphosphate (PZMA@APP) with abundant -NH_2 groups was achieved and utilized as an effective flame retardant, as a result, polyphosphazene acted a role as reactive compatibilizer, which could facilitate the dispersion of APP in epoxy resin matrix [20].

As we know, red phosphorus (RP) is a kind of commonly used flame retardant, compared to the amorphous RP, BP layered crystal could exfoliated into lamellar BP with higher thermal stability and layered structure. Thus, BP could be utilized as a novel flame-retardant additive to improve the flame retardant properties of polymers. BP plays a role to catalyze char formation and capture the free radicals; its peculiar lamellar structure can also play as a physical barrier to insulate heat and oxygen during combustion. Similar to graphene, introducing BP nanosheets into the polymer matrix exists difficulties with poor dispersion and compatibility, it is necessary to functionalize this layered material through combining with the organic compounds, the resulting inorganic–organic nanohybrids may strengthen the compatibility and interfacial interaction between the layer material and polymer matrix, thereby improving the mechanical property and flame retardant efficiency of polymer nanocomposites. Based on this, we emphasize the typical crosslinked polymers, polyphosphazene (PZN), combined with layered BP through covalently

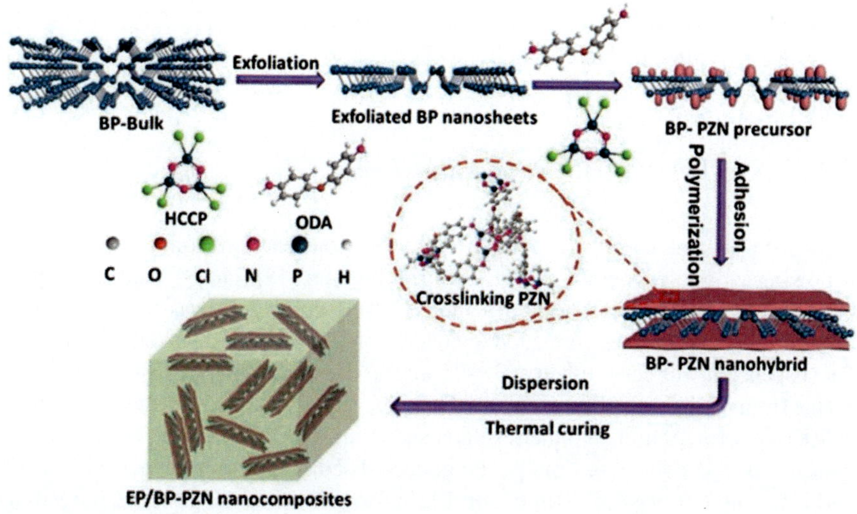

Fig. 2.1 Illustration of the fabrication process of the BP-PZN nanohybrid and EP/BP-PZN nanocomposites

grafting or electrostatic adsorption, thus, PZN plays as a promising candidate to offer improve air stability and dispersibility for BP resulting in optimized nanocomposites with synergistic effect because of the superior thermal, flame retardant, water resistance and mechanical properties of PZN. In this chapter, we developed a one-pot method to combine BP and PZN as nanohybrid, defined as BP-PZN. Phosphazene was polymerized directly in a few-layer BP suspension, resulting in an inorganic–organic nanohybrid with few-layer BP nanosheets highly encapsulated by PZN. The fabrication process of the BP-PZN nanohybrid and EP/BP-PZN nanocomposites is shown in Fig. 2.1. The resultant polyphosphazene functionalized BP (BP-PZN) with the active –NH_2 groups on the surface is beneficial to greatly increase the interfacial interaction between BP-PZN and epoxy resin (EP). The effect of PZN integrated with the BP on the flame retardant property and smoke suppression performance of EP were investigated.

2.2 Experimental Section

2.2.1 Raw Materials

4,4′-diaminodiphenyl ether (ODA), 4,4′-diaminodiphenylmethane (DDM) and hexachlorocyclotriphosphazene (HCCP), Tin (Sn 99.9%) powder and Tin (IV) iodide (99.9%) were purchased from Aladdin Industrial Corporation (China) and used without further purification. Other chemicals, red phosphorus powder (100 mesh,

98.9%) was purchased from Alfa aesar (China) Chemical co. LTD. EP (DGEBA, E-44) was provided by Hefei Jiangfeng Chemical Industry Co. Ltd. (China).

2.2.2 *Synthesis of BP-Bulk Nanosheets*

Bulk BP was synthesized by a phase transformation reaction from red phosphorus following a modified method reported in prior literature [8]. The BP nanosheets were prepared by a solvent exfoliation method. Bulk BP (100 mg) was first mixed with NMP (200 mL), and then sonicated in an ice bath (300 W) for 24 h. The as-prepared BP suspension was centrifuged at 3000 rpm for 10 min to separate disregarded sediments from nanosized BP nanosheets. Then, the supernatant was further centrifuged at 7000 rpm for 10 min, and the resultant sediment which contained few layered BP nanosheets was collected. Finally, the collected sediment was redispersed in NMP and centrifugally washed 3 times for 10 min to get rid of the residual ultra-small BP nanoparticles, and finally dried under vacuum. The dried sample was sealed in a plastic culture dish before tests, named as BP-Bulk nanosheets.

2.2.3 *Synthesis of BP-PZN Nanosheets*

BP nanosheets were dispersed in acetonitrile (0.5 mg/mL, 500 mL) by ultrasonication. ODA (140 mg) and TEA (200 mg) were then added to the suspension, respectively. Then the solution of HCCP (75 mg) in 5 mL of acetonitrile was added dropwise to the system within 30 min. The mixture was controlled at room temperture with ultrasonication (53 kHz) for 2 h, then the reaction system moved into an oil-bath and then heated to 75 °C, and then maintained for additional 24 h. After completion of the reaction, the resulting BP-PZN were collected by centrifugation at 8500 rpm for 10 min and washed with ethanol (20 mL $\times$ 3), and then dried under vacuum at 60 °C. All of above reaction process have been carried out in an inert atmosphere.

2.2.4 *Preparation of EP/BP-PZN Nanocomposites*

Preparation procedure of EP nanocomposites with 2.0 wt% BP-PZN loading illustrated below: 0.9 g of BP-PZN was dispersed in 40 mL of acetone with assistance of ultrasonication for 30 min. Follow by the corresponding 36.2 g of epoxy resin was added into the above system under sonication with mechanical stirring for 1 h. Subsequently, the redundant acetone was removed completely in a vacuum oven at 75 °C for 4 h. After that, 7.9 g of DDM was melt and poured into above blends by a rapid stirring for 1 min. Then, the resin sample was cured at 100 °C for 2 h and

150 °C for 2 h, respectively. After the curing process finished, the EP/BP-PZN2.0 sample was permitted to cool to room temperature. A similar process was utilized to prepare pure EP, EP/BP-PZN0.5 (0.5 wt%), EP/BP-PZN1.0 (1.0 wt%) and EP/BP-Bulk2.0 (2.0 wt%) composites except the type of nanoadditives.

2.2.5 Characterization

Fourier transform infrared (FTIR) spectra were performed by using a Nicolet 6700 spectrometer (Nicolet Instrument Co., USA). The samples were mixed with KBr powders and pressed into tablets before characterization. X-ray diffraction (XRD) was employed on an X-ray diffractometer (Rigaku Co., Japan), using Cu K_α radiation ($\lambda = 0.15418$ nm), at a scanning rate of 4° min^{-1}. Raman spectra were carried out on a Laser microRaman spectrometer (Jobin Yvon Co., Ltd., France) with an argon laser of 514.5 nm to study the structure components of BP-PZN and the residual char of EP composites. Transmission electron microscopy (TEM) (JEM-2100F, Japan Electron Optics Laboratory Co., Ltd., Japan) was employed to study the morphology of BP-PZN. Atomic force microscopy (AFM) was carried out using a Dimension Icon (Bruker Nano Inc.) to characterize the size and thickness of exfoliated BP nanosheets. X-Ray photoelectron spectroscopy (XPS) was conducted by using a VG ESCALAB MK-II electron spectrometer (V.G. Scientific Ltd., UK). The excitation source was an Al K_α ray at 1486.6 eV. Thermogravimetric analysis (TGA) was performed by using a Q5000 thermo-analyzer instrument (TA Instruments Inc., USA) under nitrogen atmosphere, at a linear heating rate of 20 °C min^{-1} from 50 to 800 °C. Thermogravimetric analysis/infrared spectrometry (TG-IR) was carried out on a TGA Q5000 thermo-analyzer instrument combined with a Nicolet 6700 spectrometer. Cone calorimeter test (Fire Testing Technology, UK) was performed to study the fire performance of EP composites according to the standard of ASTM E1354/ISO 5660. The combustion properties were evaluated by a microscale combustion calorimeter (MCC) with an instrument model of Govmark MCC-2 (The Govmark Organization, Farmingdale, NY). Approximately 2–5 mg of sample was heated in an inert gas atmosphere at a heating rate of 1 K s^{-1}, then the pyrolysis volatiles with carrier gases (nitrogen of 80 mL min^{-1}; oxygen of 20 mL min^{-1}) were mixed and burned at 900 °C in a combustion chamber. Scanning electron microscopy (SEM, JEOL JSM-6700) was conducted to study fracture surface and microstructures of the char residues for EP composites. Dynamic mechanical analysis (DMA) was conducted with the PerkinElmer Pyris Diamond DMA from room temperature to 250 °C at a linear heating rate of 5 °C/min, at a frequency of 1 Hz in the tensile configuration.

2.3 Results and Discussion

2.3.1 *Morphology and Structure Characterization of BP-Bulk and BP-PZN*

To unveil the structure and composition of BP-PZN compared to the BP-Bulk, a series of morphological and spectroscopic characterizations were conducted. The FTIR spectrum of BP-PZN is shown in Fig. 2.2, there exist several characteristic peaks correspond to the inherent structure of BP crystal which both observed in IR spectrum of BP-Bulk, otherwise, some new peaks can be seen as follow, two apparent peaks at 1209 and 868 cm^{-1} are assigned to the P=N and P–N characteristic absorption of the polyphosphazene, respectively [17]. The intense peak at 1157 cm^{-1} corresponds to the stretching vibration of the ether bond. The typical peaks at 1630 and 1500 cm^{-1} are associated to the stretching vibration of the C=C group in the phenylene of ODA, are all observed in the spectra of BP-PZN.

The XRD patterns of BP-Bulk and BP-PZN powders are well indexed to the orthorhombic phase of BP, as shown in Fig. 2.3a. For BP-Bulk, its XRD pattern shows several representative diffraction peaks at 17.0°, 26.7°, 34.3°, 52.5°, 56.2° and 56.8° corresponding to the (020), (021), (040), (060), (151) and (061) planes, respectively, which are typical layered planes of BP-Bulk [12, 21]. For BP-PZN, it shows all of the diffraction peaks in BP-Bulk and its peaks are weaker, due to the polyphosphazene polymer is coating on the surface of BP nanosheets. The structural integrity of the BP-Bulk and BP-PZN was further confirmed by analyzing the Raman spectra (Fig. 2.3b). Three representative vibrational modes of BP-Bulk are assigned to the peaks of A_g^1 at 361.6 cm^{-1}, B_{2g} at 439.7 cm^{-1}, and A_g^2 at 466.2 cm^{-1} [22]. The intensity ratio of A_g^1/A_g^2 is approximately 0.7, this demonstrates that there is no oxidation of the BP nanosheets [23]. These peaks are also detected for BP-PZN sample, demonstrating the retention of the vibrational structure after polyphosphazene functionalization of

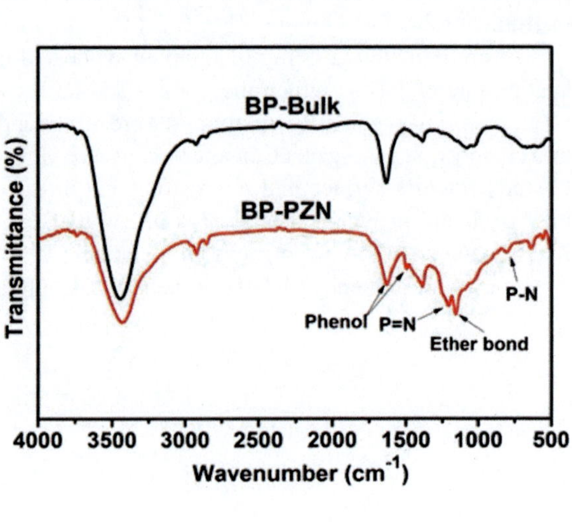

Fig. 2.2 FT-IR spectra of BP-bulk and BP-PZN nanosheets

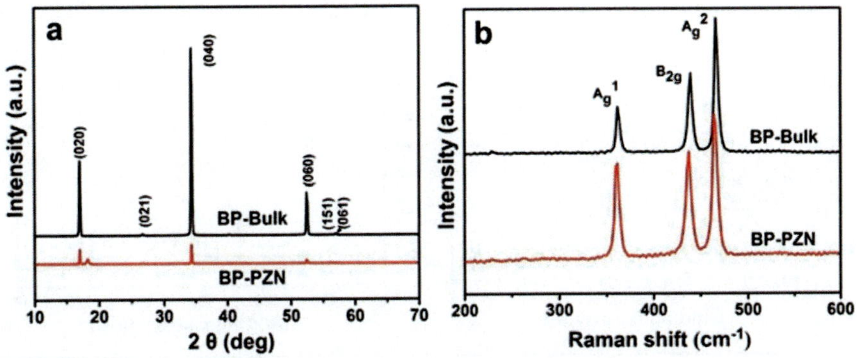

Fig. 2.3 **a** XRD patterns and **b** Raman spectra of BP-Bulk and BP-PZN nanosheets

BP. However, Raman spectrum of the BP nanosheets shows more intense peaks. It can be observed that the three Raman peaks of the BP-PZN nanosheets red shift slightly (about 2 cm^{-1}) compared to those of the BP-Bulk. These results give the further evidence that the thickness of nanolayer increases in BP nanosheets after coating with polyphosphazene polymer [24].

Thermogravimetric analysis (TGA) was then conducted under nitrogen so as to confirm the mass loss process and formation of polyphosphazene in BP-PZN. As shown in Fig. 2.4, bulk BP exhibits a one-step mass loss between 450 and 550 °C. However, BP-PZN has two steps mass loss until 800 °C. The first mass loss step is as same as the degradation of bulk BP, and the second step between 500 and 600 °C, due to the decomposition of polyphosphazene cross-linking polymers.

The valence states and chemical composition of the BP-Bulk and BP-PZN were evaluated by XPS analysis, as shown in Fig. 2.5. In the survey spectra (Fig. 2.5a), the signal peaks of C, O, N, P elements can be obviously distinguished, in the spectrum of BP-PZN, and the elements content of C, O is increased, which identify

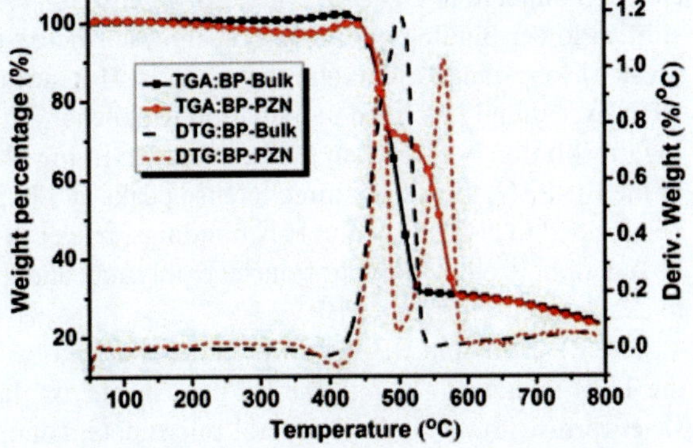

Fig. 2.4 TGA and DTG curves of BP-bulk and BP-PZN nanosheets

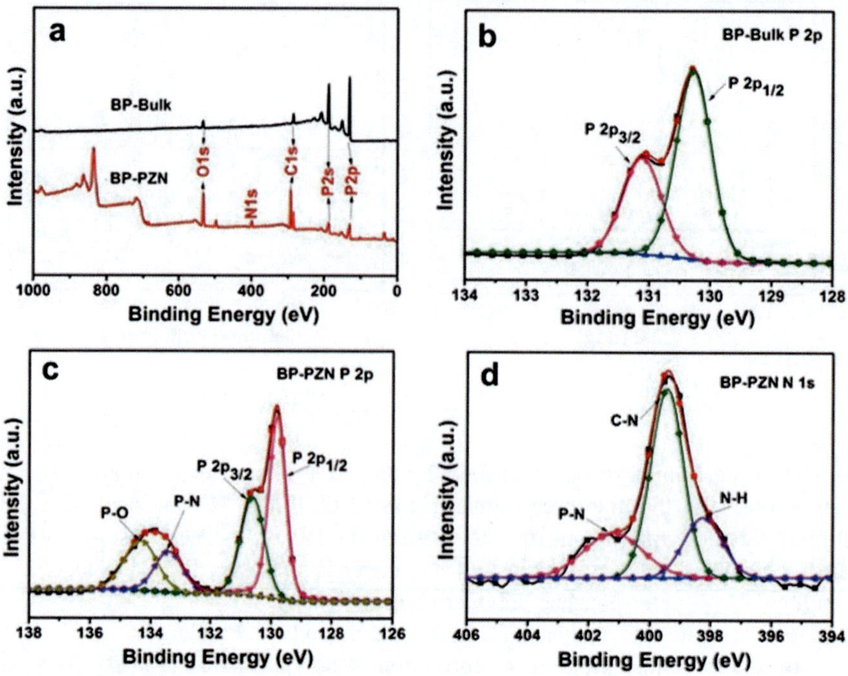

Fig. 2.5 **a** XPS survey spectra of BP-Bulk and BP-PZN; **b**, **c** High-resolution P 2p XPS spectra of BP-bulk and BP-PZN; **d** high-resolution N 1 s XPS spectra of BP-PZN

that the successful formation of BP-PZN nanosheets. Figure 2.5b presents high-resolution P2p XPS spectrum of BP-Bulk, are deconvoluted into two peaks at 130.2 and 131.1 eV correspond to $P2p_{3/2}$ and $P2p_{1/2}$ of P-P bonds, respectively. Figure 2.5c–d present high-resolution P2p and N1s XPS spectra of BP-PZN. For P2p peaks of BP-PZN, which can be deconvoluted into four peaks at 129.8, 130.6, 133.4, and 134.3 eV correspond to $P2p_{3/2}$ and $P2p_{1/2}$ of P-P bonds, P-N and P-O bonds, respectively. Compared to BP-Bulk, the $P2p_{3/2}$ and $P2p_{1/2}$ signals of BP-PZN are slightly shift to lower binding energies, revealing an increase of electron concentration, a break of long-range order, and the oxidation for surface P atoms of BP nanosheets. BP is exceedingly inclined to oxidation in open air. These results are also testified by high-resolution N1s XPS analysis, as shown in Fig. 2.5d. In the N1s XPS spectrum of the BP-PZN, there exist three intense peaks at 398.3, 399.5 and 402.1 eV that can be ascribed to N–H, C-N and P-N bondings, respectively, revealing the bulk BP can be functionalized by polyphosphazene polymer, and the polyphosphazene are prefer to attach on the surface of BP.

The BP-Bulk and BP-PZN were characterized by TEM and AFM. As shown in the TEM image (Fig. 2.6a), the BP nanosheets exfoliated from BP-Bulk is in large sheet form with a length of several micrometers and well maintained crystallinity. In Fig. 2.6b, c, the high-resolution TEM (HRTEM) image and the selected area

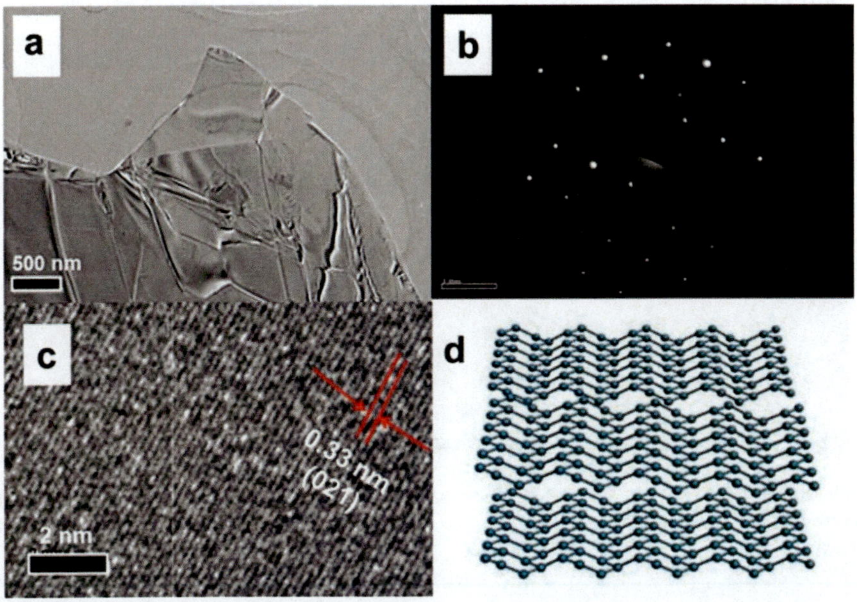

Fig. 2.6 **a** TEM image of the BP nanosheets and the corresponding **b** SAED pattern and **c** HRTEM image of BP nanosheets; **d** the atomic structure model of monolayer and folded-bilayer BP nanosheets

electron diffraction (SAED) pattern show the orthorhombic crystalline structure of the BP nanosheets. The lattice fringes of 0.33 nm is assigned to the (021) plane of BP, which is in accordance with the results in XRD analysis [25]. Figure 2.6d shows the atomic structure model of monolayer and folded-layer BP nanosheets.

A magnified AFM image and the height profiles of single BP nanosheets are shown in Fig. 2.7a, b. BP nanosheets have smooth surfaces with an average thick-

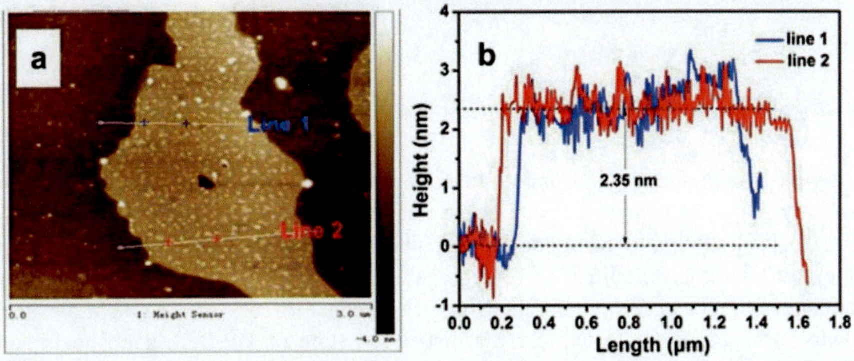

Fig. 2.7 **a** AFM image of BP nanosheets and **b** the corresponding height profiles taken along the lines marked in (**a**)

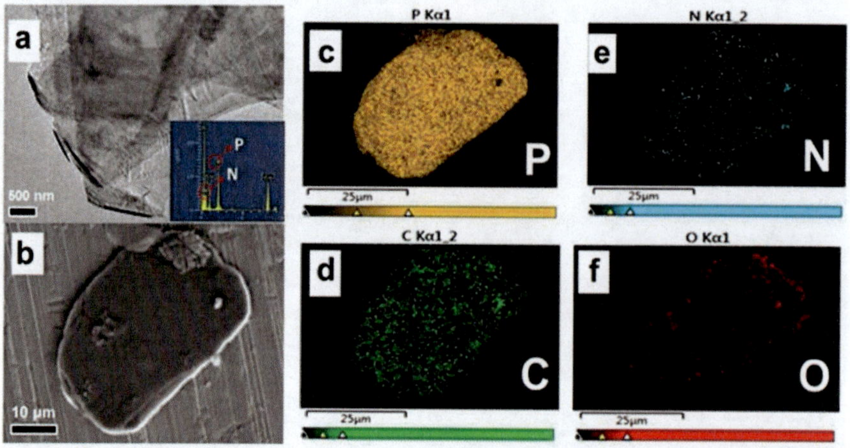

Fig. 2.8 **a** TEM image of the BP-PZN nanosheets; **b** SEM image of BP-PZN nanosheets and corresponding elemental mapping images of **c** phosphorus (P), **d** carbon (C), **e** nitrogen (N) and **f** oxygen (O)

ness of 2.35 nm. Normally, the thickness of single-layer phosphorene is considered as 0.53 nm [24]. Thereby, the thickness of 2.35 nm observed in Fig. 2.7b, corresponds to 4–5 layers of P atoms. Such few-layer BP could be defined as few-layer black phosphorene materials. Additionally, the polyphosphazene polymers as an extra phase which grow on the BP nanosheets can be clearly observed in Fig. 2.8a. After functionalized with polyphosphazene polymer, the exfoliated BP-PZN maintains the well nanosheet morphology. Moreover, as shown in the inset of Fig. 2.8a, the energy dispersive spectrometer (EDS) results confirmed the elements distribution in the BP-PZN nanosheets. Figure 2.8b–f shows the SEM image of BP-PZN nanosheets (b) and corresponding elemental mapping images (c–f). The successful formation of BP-PZN nanosheets was further verified by elemental mapping images of phosphorus (P), nitrogen (N), oxygen (O), and carbon (C).

2.3.2 Characterization of EP/BP-PZN Nanocomposites

As well-known, the well dispersion state of nanofillers in polymer matrix is beneficial to improving the mechanical and flame retardant properties of polymer nanocomposites [26]. Owing to the strong Van der Waals interaction between BP nanolayers, they are extremely inclined to re-aggregate in polymer matrix, which have deteriorating effects on the comprehensive performance of polymer nanocomposites. Thus, it is important to explore the dispersion state of BP-Bulk nanosheets and BP-PZN nanosheets in EP matrix. The morphologies of EP/BP-Bulk2.0 and EP/BP-PZN2.0 nanocomposites were investigated by TEM (Fig. 2.9a, b). From Fig. 2.9a, re-aggregation and non-uniform dispersion of the BP-Bulk nanosheets can be observed

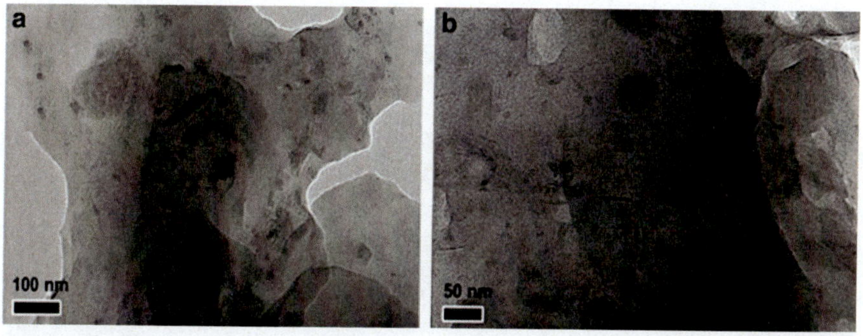

Fig. 2.9 TEM images of the **a** EP/BP-Bulk2.0 and **b** EP/BP-PZN2.0 nanosheets

for EP/BP-Bulk2.0 sample. As a contrast, from the microstructures and intercalating state of EP nanocomposites with 2.0 wt % BP-PZN, the homogeneous dispersion of BP-PZN are observed in EP matrix (Fig. 2.9b), due to the functionalization of BP with polyphosphazene polymers can facilitate the interfacial interaction and compatibility between BP nanosheets and EP matrix, thereby improve the dispersion state of this nanoadditives in EP matrix.

The SEM images of freeze-fractured surface for EP, EP/BP-Bulk2.0 and EP/BP-PZN2.0 nanocomposites are shown in Fig. 2.10a–c. As shown in Fig. 2.10a, the fractured surface of pure EP is extremely smooth and shows a no-crinkled morphology. But for EP/BP-Bulk2.0 nanocomposites (Fig. 2.10b), it can be seen that the fracture surface is rough with small-crinkled morphology, implying interface interaction between BP-Bulk nanosheets and EP matrix. Furthermore, EP/BP-PZN2.0 nanocomposite exhibits a more rough fractured surface with large-crinkled morphology, than pure EP and EP/BP-Bulk2.0 (Fig. 2.10c). This may be due to the interpretations that functionalization of BP with polyphosphazene polymers leading to the improved interfacial adhesion between BP-PZN and EP matrix.

2.3.3 *Thermal and Mechanical Properties of EP/BP-PZN Nanocomposites*

The effect of BP-PZN nanosheet on the thermal property of EP nanocomposites was evaluated by TGA analysis under nitrogen (N_2), and the detail data are shown in Table 2.1. Upon the Fig. 2.11a, EP/BP-Bulk and EP/BP-PZN nanocomposites exhibit a one-step weight loss process in the region of 300–500 oC, which have similar pyrolysis behavior to pure EP. The onset degradation temperature of the EP/BP-PZN nanocomposites decreased slightly, however, with increasing of the BP-PZN content, their char residues at 800 °C gradually increased significantly, higher than that of neat EP, owing to the catalytic charring effect of polyphosphazene and the superior thermal stability of BP nanosheets [27]. Moreover, the initial decomposition temperature of

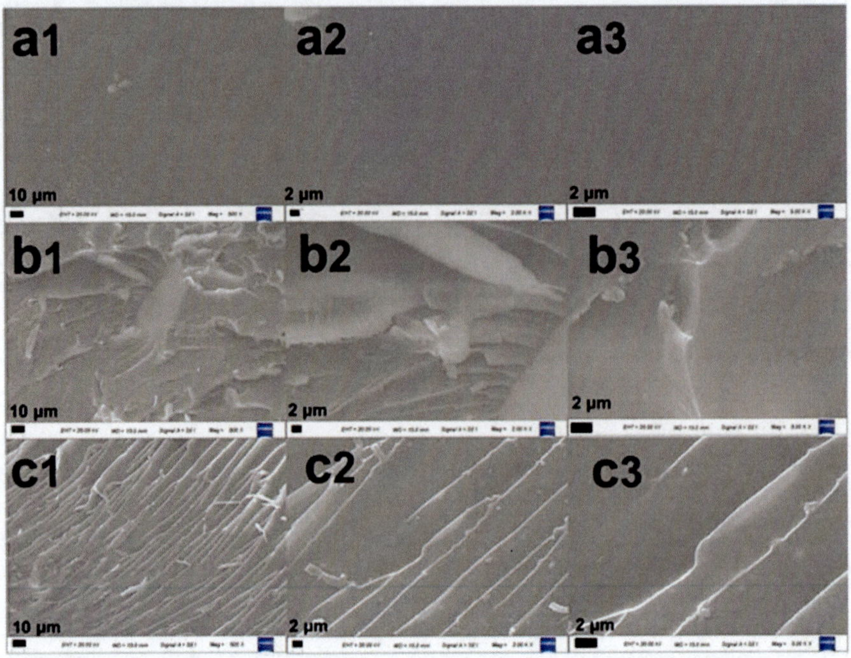

Fig. 2.10 SEM images of the (a1-3) pure EP, (b1-3) EP/BP-Bulk2.0 and (c1-3) EP/BP-PZN2.0 nanocomposites in different magnification

Table 2.1 TGA data for EP and its nanocomposites in nitrogen

Sample	$T_{-5\%}$ (°C)	$T_{-50\%}$ (°C)	T_{max} (°C)	Residue at 800 °C (wt%)
EP	375.3	410.1	392.5	12.8
EP/BP-PZN0.5	361.2	408.3	396.9	16.4
EP/BP-PZN1.0	361.0	407.0	395.5	16.2
EP/BP-PZN2.0	355.5	408.6	398.2	19.4
EP/BP-Bulk2.0	371.9	410.8	396.7	15.5

EP/BP-Bulk2.0 sample is close to pure EP, with a char yield of 15.6% at 800 °C, lower than char yield (19.5%) of EP/BP-PZN2.0 after combustion with the same loading, because of the earlier degradation of EP nancomposites resulting from the catalytic charring effect of polyphophazene. From the derivative thermogravimetric analysis (DTG) curves (Fig. 2.11b), the maximum mass loss rates of EP/BP-Bulk and EP/BP-PZN naocomposites are much lower than that of pure EP. BP nanosheets are inorganic part as a physical barrier which can inhibit oxygen exchange and diffusion of pyrolysis products, polyphosphazenes are organic part which promote the char formation, thus improving thermal resistance of the nanocomposites against degradation.

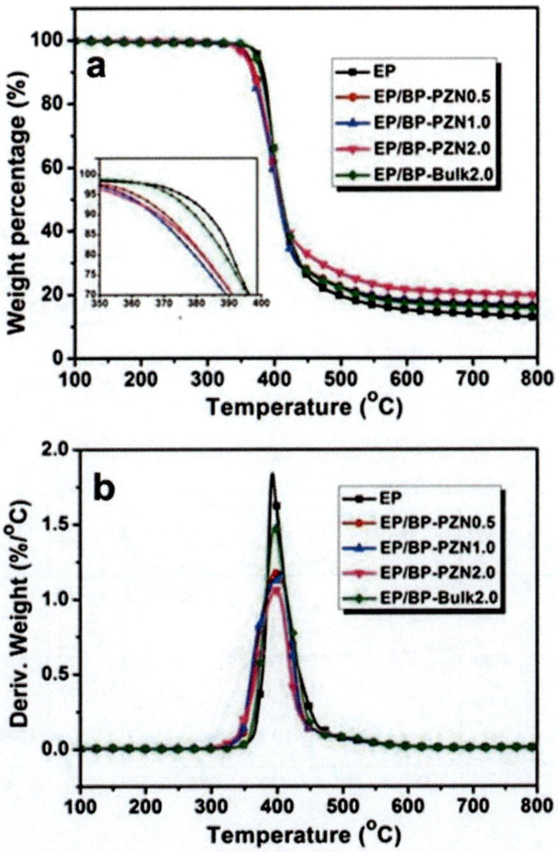

Fig. 2.11 **a** TGA curves and **b** DTG curves of the pure EP and its nanocomposites with different BP-PZN contents

The effect of BP-PZN nanosheet on the thermal property of EP nanocomposites was evaluated by TGA analysis under air. Upon the Fig. 2.12, EP/BP-Bulk and EP/BP-PZN nanocomposites exhibit a three-step weight loss process in the region of 250–600 oC, which have similar pyrolysis behavior to pure EP. With increasing of the BP-PZN content, their char residues at 800 °C gradually increased significantly, higher than that of neat EP, owing to the catalytic charring effect of polyphosphazene and the superior thermal stability of BP nanosheets.

The mechanical and thermal properties were studied by DMA to verify these performances were improved after introducing BP-Bulk and BP-PZN nanosheets into EP matrix. The storage modulus and tan δ of the BP-Bulk and BP-PZN nanocomposites as a function of temperature are shown in Fig. 2.13a, b. The pure EP exhibits a storage modulus at room temperature of approximately 23,253.7 MPa. The storage modulus values at room temperature of EP/BP-PZN nanocomposites are significantly improved by 61.5%, 128.9%, 40.0%, respectively, with increasing the content of BP-PZN nanosheets from 0.5 to 2.0 wt%, compared to pure EP. These satisfied improvements in the storage modulus are primarily attributed to the high stiffness of

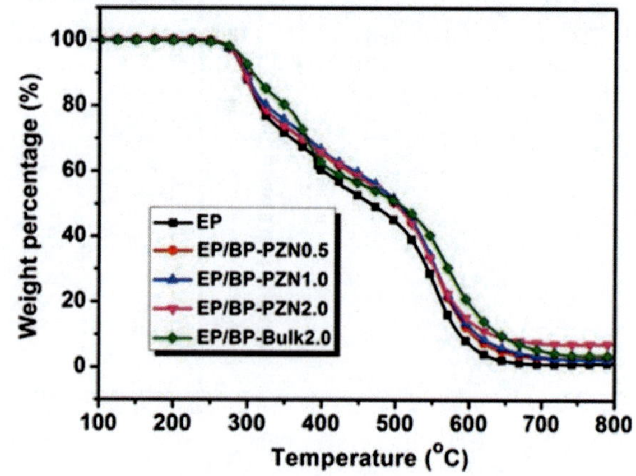

Fig. 2.12 TGA curves of the pure EP and its nanocomposites with different BP-PZN contents under air atmosphere

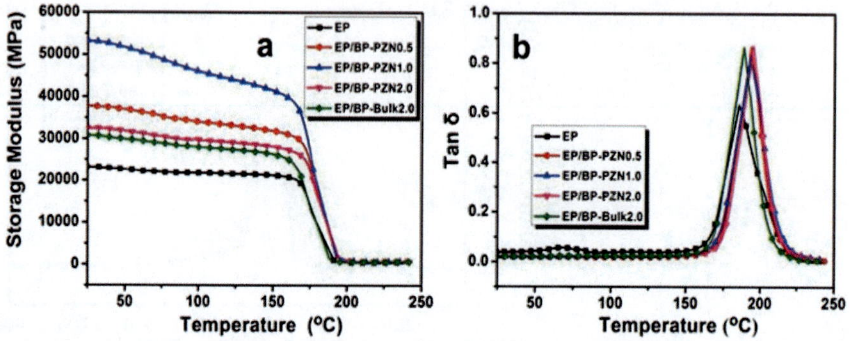

Fig. 2.13 **a** Storage modulus (E') curves and **b** Tan δ curves of the pure EP and its nanocomposites with different BP-PZN contents as a function of temperature

BP nanosheets and enhanced interfacial interaction between the BP-PZN and EP. In particular, The DMA data for EP/BP-Bulk2.0 sample exhibits a storage modulus at room temperature of 30,814.8 MPa, which is much lower than that of EP/BP-PZN2.0. The EP/BP-PZN2.0 show relative higher thermal stability in terms of modulus than that of EP/BP-Bulk2.0, owing to that the inherent characteristics and structural organization of BP nanosheets and crosslinking polyphosphazene, combined with tight mixing EP. The peak value of tan δ can be ascribed to glass transition temperature (T_g) of the EP/BP-PZN nanocomposites (Fig. 2.13b). In detail, with the introducing of various loadings of BP-PZN into EP matrix, the peaks of tan δ are slightly shift to higher temperatures. The increase of glass transition temperature is attributed to the enhancement of the compatibility and interfacial interaction between the BP nanosheets and EP molecular chains, thus improving the thermal and mechanical performances of EP/BP-PZN nanocomposites.

2.3.4 *Flame Retardancy of EP/BP-PZN Naocomposites*

The fire retardant properties of EP nanocomposites were investigated by cone calorimeter test which can imitate real combustion environment of polymer materials. The heat release rate (HRR) and total heat release (THR) versus time curves of EP nanocomposites are shown in Fig. 2.14a, b. Pure EP burns dramatically with high peak heat release rate (PHRR) and THR values of 2116.0 kW/m^2 and 167.1 MJ/m^2, respectively. With increasing BP-PZN content from 0.5 to 2.0 wt%, the PHRR values of the EP/BP-PZN nanocomposites decreased by 23.7% to 59.4%; besides, the THR values of the EP/BP-PZN nanocomposites decreased by 28.3% to 63.6%. Particularly, the introduction of 2 wt% BP-PZN in EP leads to a maximum 59.4% decrease in PHRR, a maximum 63.6% decrease in THR during combustion, implying the increased flame retardant properties of EP/BP-PZN nanocomposites with the increasing of BP-PZN content. The PHRR and THR values for the EP/BP-Bulk2.0 nanocomposites are reduced by 48.9% and 43.6%, respectively, the reduction scale are lower than those of EP/BP-PZN2.0 nanocomposites.

Figure 2.15a, b present the smoke production rate (SPR) and total smoke release (TSR) versus time curves of EP nanocomposites. Pure EP can easily release heavy smoke during combustion with higher SPR and TSR values due to its multi-aromatic structure. However, the *SPR* and TSR values of the EP/BP-PZN nanocomposites with different loadings are significantly reduced compared to pure EP, and EP/BP-PZN2.0 exhibits the lower *SPR* and TSR values, than that of EP/BP-Bulk2.0. The significant reductions in the fire hazards of EP composites are attributed to the physical barrier effect of BP nanosheets and cooperative catalytic charring effect between the PZN and BP, thereby retarding the diffusion of pyrolysis products and forbid the inner unburned materials exposed to fire during combustion.

To further evaluate the thermal degradation behavior of EP/BP-Bulk and EP/BP-PZN nanocomposites, the volatile products diffused from EP, EP/BP-Bulk2.0 and EP/BP-PZN2.0 were investigated by TG-FTIR technique. Figure 2.16 shows the

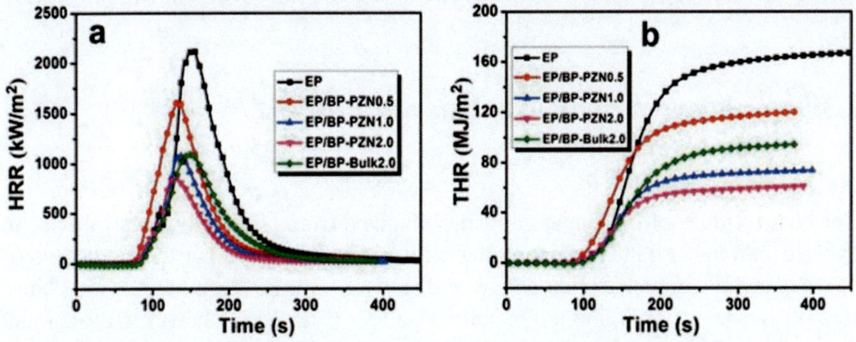

Fig. 2.14 **a** HRR and **b** THR versus time curves of the pure EP and its nanocomposites with different BP-PZN contents from cone tests

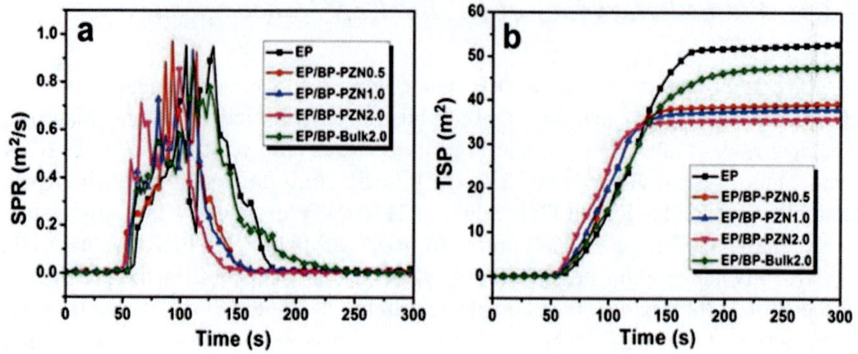

Fig. 2.15 **a** SPR and **b** TSR versus time curves of the pure EP and its nanocomposites with different BP-PZN contents from cone tests

intensities of the characteristic pyrolysis products versus time of EP nanocomposites. Several representative peaks for the evolved gaseous products of EP nanocomposites as follow: the peak at 1510 cm^{-1} is assigned to absorption of aromatic compounds and 1740 cm^{-1} is attributed to absorption of carbonyl compounds; the peak at 2190 cm^{-1} is assigned to absorption of carbon monoxide (CO); the peak at 2360 cm^{-1} is ascribed to absorption of carbon dioxide (CO_2); the peak at 2930 cm^{-1} corresponds to absorption of hydrocarbons [28]. With the introduction of 2.0 wt% of BP-Bulk and BP-PZN into EP, the maximum absorbance intensities of characteristic volatile products, including hydrocarbons and aromatic compounds, carbonyl compounds, CO and CO_2, are significantly decreased, lower than those of pure EP. As expected, EP/BP-PZN2.0 exhibits lowest absorbance intensity of above volatile products among pure EP and EP/BP-Bulk2.0 nanocomposites. CO is considered to be the primary toxic substance in the combustion process of EP, and the reduction of CO concentration contributes to decrease smoke toxicity; the decrease of intensities of flammable volatile products such as hydrocarbons and aromatic compounds is beneficial to reducing smoke and heat release, thereby improving the fire safety [20].

2.3.5 Condensed Phase Analysis of EP/BP-PZN Nanocomposites

The char residues of EP nanocomposites obtained from cone test were evaluated to explore reasonable flame-retardant mechanism. Figure 2.17 shows the digital photos of the external residues from top view and side view for EP and its nanocomposites. Few of cracked residual char left in pure EP sample (Fig. 2.17a), however, the amount and size of residual chars are gradually increased with the increasing the contents of BP-PZN (0.5, 1.0, 2.0 wt%) (Fig. 2.17b–d), and an exceedingly expanded and dense char layer can be observed in EP/BP-PZN2.0 nanocomposite (Fig. 2.17d), the

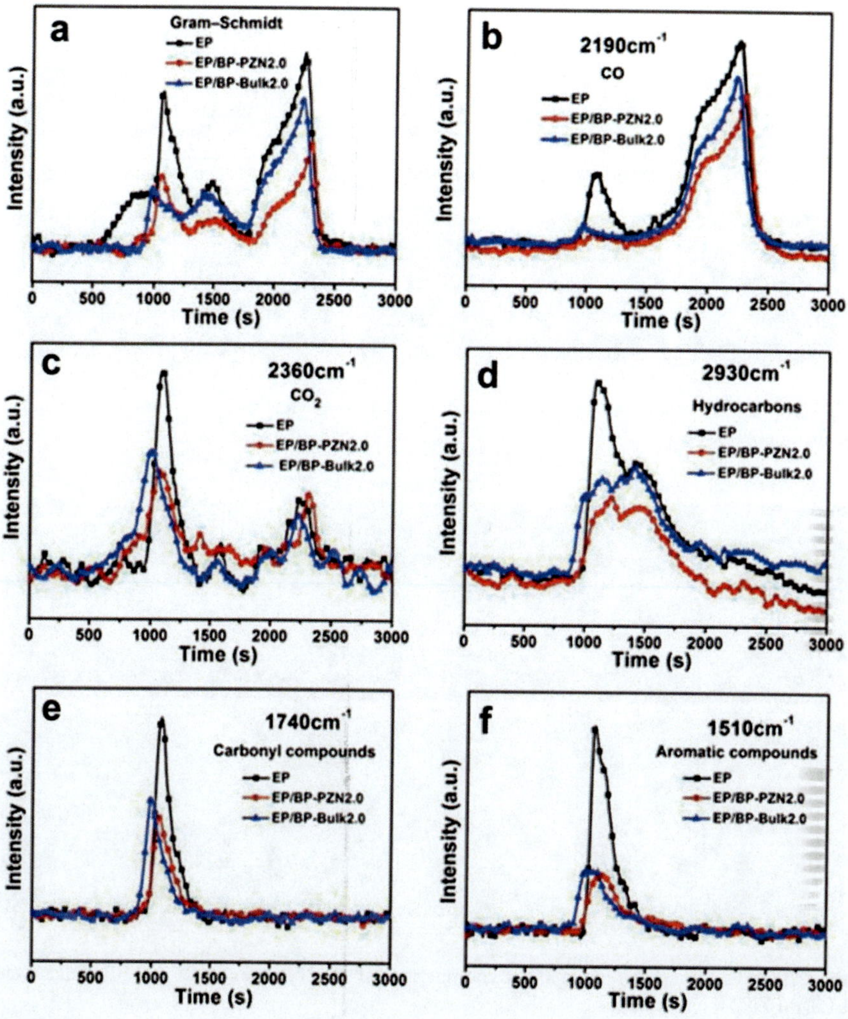

Fig. 2.16 Absorbance of pyrolysis products for EP, EP/BP-PZN2.0 and EP/BP-Bulk2.0 sample versus time: **a** total pyrolysis products; **b** CO; **c** CO_2; **d** hydrocarbons; **e** carbonyl compounds and **f** aromatic compounds

quality of this char layer is better than that of EP/BP-Bulk2.0 (Fig. 2.17e) with the same loading from combustion. As consider the EP/BP-PZN nanocomposites, the continuous, compact and expanding char layers are formed due to the physical barrier effect of BP nanosheets, intumescent effect of polyphosphazene and cooperative catalytic charring effect between the PZN and BP nanosheets.

The microstructures of external (Fig. 2.18a–c) and interior (Fig. 2.18d–f) residues for EP, EP/BP-Bulk2.0 and EP/BP-PZN2.0 nanocomposites are displayed in Fig. 2.18. SEM image of external char layer for pure EP exhibits a flaky morphology

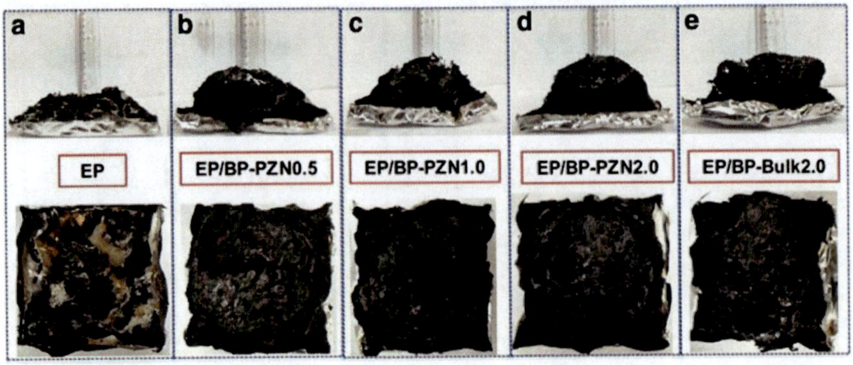

Fig. 2.17 Digital photos of the external residues from top view and side view for EP and its nanocomposites

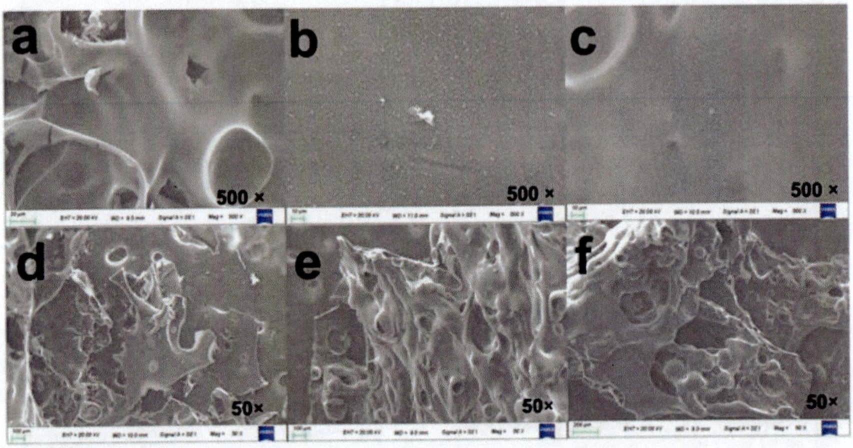

Fig. 2.18 SEM images of external **a–c** and interior **d–f** char residues for EP, EP/BP-Bulk2.0 and EP/BP-PZN2.0 nanocomposites

with big opening hole on the surface (Fig. 2.18a). As compared, for the EP/BP-Bulk2.0 sample, the external residue shows a more compact and complete surface with no holes (Fig. 2.18b). As expect, after the EP/BP-PZN2.0 (Fig. 2.18c) combustion from cone test, it can be observed that the continuous and more compact surfaces of external residue are formed. The SEM images of internal residues (Fig. 2.18d–f) further confirm that the internal char residues of EP/BP-Bulk2.0 and EP/BP-PZN2.0 are more compact than that of pure EP. Many big opening holes and fragments are observed in internal residues of pure EP, as compared, there are few holes with more condense inwall for the internal char layers of EP/BP-Bulk2.0 and EP/BP-PZN2.0 sample, due to the BP nanosheets can strengthen the quality of char layer during the combustion, this may be due to the lack of oxygen in underlying polymers during burning. As is known to all, the char layer has a denser and more cohesive surface, which is conductive to delaying the mass transfer, heat transfer and release of volatile products, thus improving the safety of fire.

The carbonized char layer with high carbonization degree is an effective barrier to prevent the internal material ignition. Raman spectroscopy was utilized to study the composition and structure of the char residue. Figure 2.19a–c show the Raman spectra of EP, EP/BP-Bulk2.0 and EP/BP-PZN2.0 nanocomposites, it can be observed that two representative peaks at 1365 and 1596 cm^{-1}, are defined as D peak and G peak, respectively. In prior works, the graphitization degree of char residue is measured by the area ratio of D peak to G peak (I_D/I_G), and lower I_D/I_G value equals to higher graphitization degree [29, 30]. In particular, the I_D/I_G value of pure EP is 2.69, whereas the EP/BP-Bulk2.0 nanocomposite presents a I_D/I_G value of 2.54, the EP/BP-PZN2.0 nanocomposite shows lower I_D/I_G value (2.52) among three samples, revealing the higher graphitization degree, due to the cooperative catalytic charring effect between polyphosphazene and few-layer BP nanosheets during EP combustion.

FTIR spectra (a) and XRD patterns (b) of the char residues for EP and EP/BP-PZN nanocomposites after cone tests are shown in Fig. 2.20. In IR spectra (Fig. 2.20a), the peaks at 736 and 1586 cm^{-1} are associated to multi-aromatic structure, which

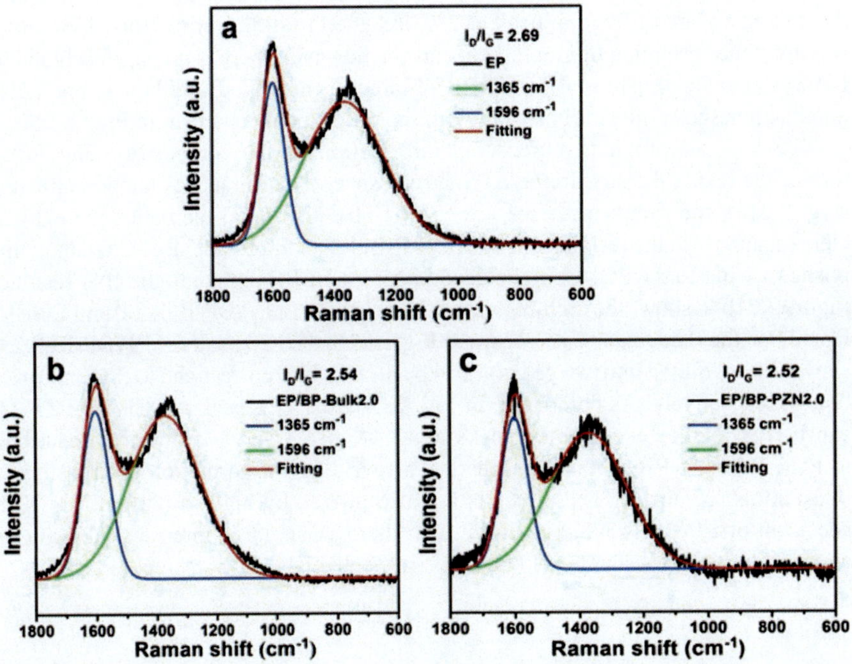

Fig. 2.19 Raman spectra of the char residues of **a** EP, **b** EP/BP-Bulk2.0 and **c** EP/BP-PZN2.0

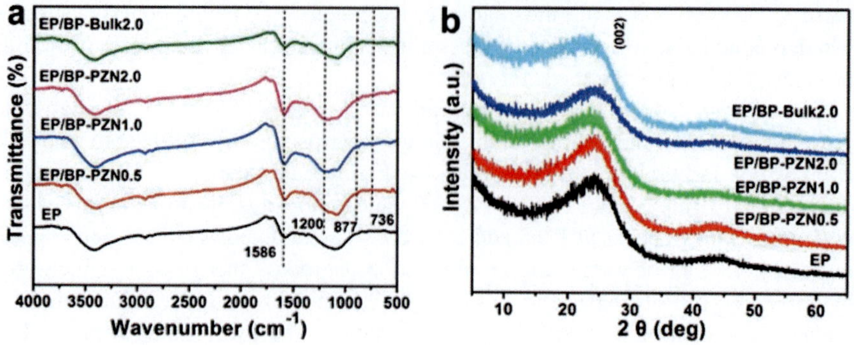

Fig. 2.20 **a** FTIR spectra and **b** XRD patterns of the residual char for EP and its nanocomposites after cone tests

can be observed in all of EP and its nanocomposites. For the residual char of EP/BP-Bulk2.0 and EP/BP-PZN2.0 nanocomposites, a weak peak at 1200 cm^{-1} appeared, corresponds to phosphoryl ($P = O$) stretching modes for the burned BP. In addition, a absorption peak at 877 cm^{-1} can also be observed, assigned to P–N typical absorption of cyclotriphosphazene, revealing the formation of crosslinked phosphorus oxynitride [31]. Figure 2.20b shows the XRD pattern of EP and its nanocomposites, there exist a wide diffraction peak at 23°, assigned to (002) diffraction of graphite, revealing the formation of graphitized carbon, however, there is no peak assigned to BP crystal, indicating most BP and BP-PZN are fully degraded after high temperature combustion, the results is consistent with the proposed mechanism in Fig. 2.22.

XPS analysis provides more detail information on the composition and structure of the residual chars. In the XPS survey spectra of EP and its nanocomposites (Fig. 2.21a), the samples surface are consist of C, N and O elements, the extra P element appear in the residual char of EP/BP-Bulk2.0 and EP/BP-PZN2.0 nanocomposites, owing to the PO_x from the decomposition of BP remain in the char residue. Figure 2.21b–d shows the high-resolution XPS spectra of EP/BP-Bulk2.0 and EP/BP-PZN2.0 in the P 2p regions. As shown in Fig. 2.21b, the P2p peaks of EP/BP-Bulk2.0 can be deconvoluted into two peaks at 134.3 and 135.4 eV correspond to P-O and P_2O_5 bonds, respectively. As compared, in Fig. 2.21c, the P2p peaks of EP/BP-PZN2.0 can be deconvoluted into three peaks at 133.4, 134.3 and 135.4 eV, are ascribed to P-N, P-O and P_2O_5 bonds, respectively, revealing the formation of crosslinked phosphorus oxynitride. These results are also proved by high-resolution N1s XPS spectrum of EP/BP-PZN2.0 in Fig. 2.21d, there exists three intense peaks can be assigned to N–H, C-N and P-N bondings, respectively.

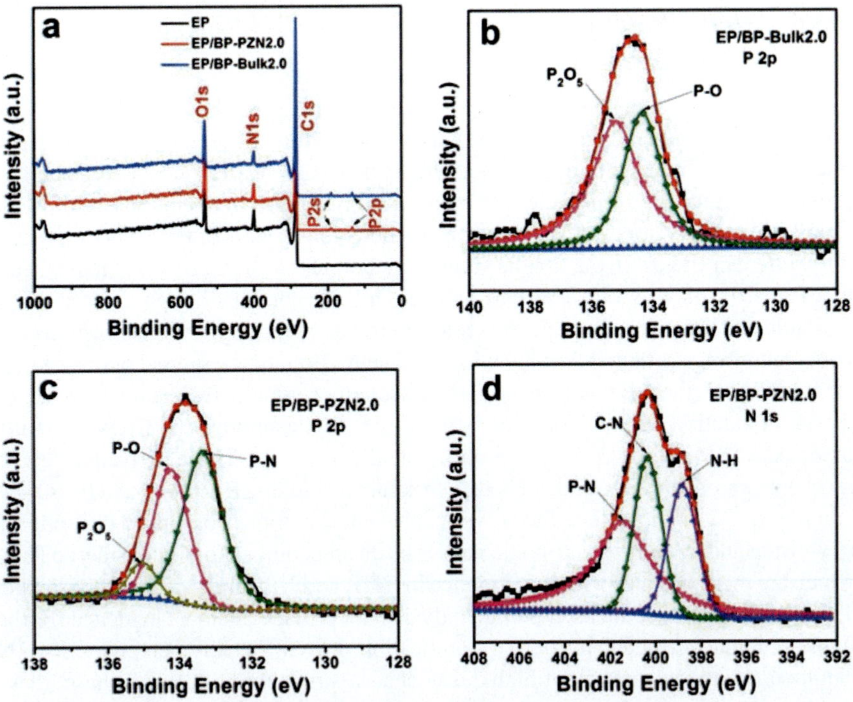

Fig. 2.21 **a** XPS survey spectra of the residual char for EP and its nanocomposites after cone tests; High-resolution P 2p XPS spectra of **b** EP/BP-Bulk2.0 and **c** EP/BP-PZN2.0; **d** High-resolution N 1 s XPS spectra of EP/BP-PZN2.0

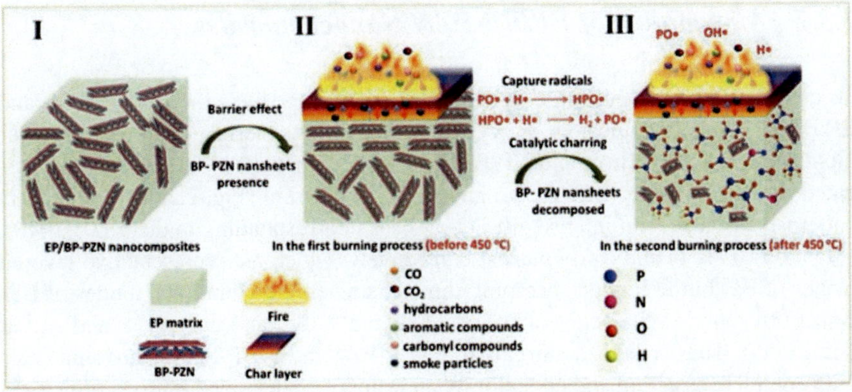

Fig. 2.22 Schematic illustration of flame-retardant mechanism for BP-PZN in flaming EP composites

2.3.6 Flame Retardant Mechanism of EP/BP-PZN Nanocomposites

Based on the above analysis for combustion behavior in gaseous and condensate phases, a possible flame-retardant mechanism of EP/BP-PZN nanocomposites is proposed in Fig. 2.22. Take the maximum decomposition temperature of BP in air (~450 °C) as the watershed, we divide the combustion process of EP nanocomposites into two main stages. In the first burning stage before ~ 450 °C, BP-PZN nanosheets in EP matrix play as physical barrier to retard the combustible gases containing aromatic and hydrocarbon fragments release from pyrolysis epoxy to the combustion zone, inhibiting sustained flaming with heat generation. In the second burning stage after ~ 450 °C, the BP nanosheets start to decompose gradually. As the increase of thermal oxidation degradation time, similar to red phosphorus, the BP was mainly oxidized into various PO_x and phosphoric acid derivatives. These derivatives could react with epoxy to generate more stable structures including P—O—C and P—O—P complexes [32]. For another, the polyphosphazene can form crosslinked phosphorus oxynitride and carbonized aromatic networks during combustion is considered [27], which are in accordance with the evidence in FTIR and XPS analysis for char residue. These gaseous and condensed phase analysis data provide adamant evidence for the flame retardant mechanism: during combustion process of polymers, the BP-PZN can facilitate the generation of high dense char layers in the condensed phase, these dense char layer retard mass and heat transfer and inhibit further erosion of the underlying polymeric substrate by heat flow in a burning zone.

2.3.7 Air Stability of EP/BP-PZN Nanocomposites

To evaluate the air stability of EP/BP-PZN nanocomposites, the EP/BP-Bulk and EP/BP-PZN nanocomposites were exposed to atmospheric conditions for four months. The EP/BP-Bulk and EP/BP-PZN nanocomposites were further investigated by XRD pattern (Fig. 2.23a) and Raman spectrum (Fig. 2.23b). Their XRD patterns exhibit several intense diffraction peaks corresponding to the (020), (040), (111), (060) (151) and (061) planes, respectively, which are representative layered planes of BP. In the Raman spectrum, three characteristic vibrational modes of BP-Bulk correspond to the peaks of A_g^1 at 361.0 cm^{-1}, B_{2g} at 439.1 cm^{-1}, and A_g^2 at 465.7 cm^{-1}. These results confirm that the BP-PZN in EP/BP-PZN nanocomposites with complete structure and crystallinity and cannot be degraded even on prolonged exposure to the air more than four month. The uniformly distributed crystalline BP-Bulk and BP-PZN nanosheets in the EP/BP-Bulk and EP/BP-PZN nanocomposites, respectively, can be clearly seen by TEM observations of the ultrathin sections obtained from (2.23c) EP/BP-Bulk2.0 and (2.23d) EP/BP-PZN2.0 nanocomposites

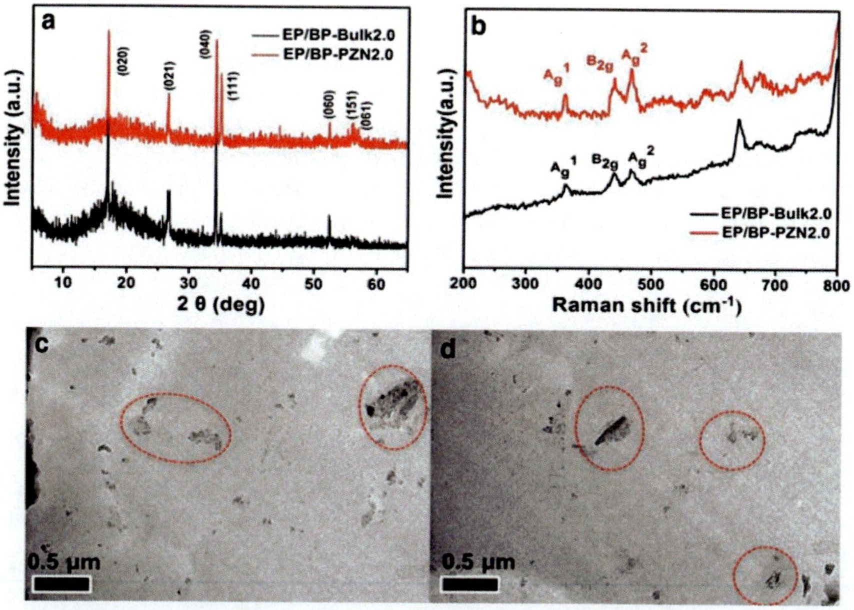

Fig. 2.23 The corresponding **a** XRD pattern and **b** Raman spectra of EP/BP-Bulk2.0 and EP/BP-PZN2.0 nanocomposites after exposure to ambient conditions for four months; The TEM observations of the ultrathin sections obtained from **c** EP/BP-Bulk2.0 and **d** EP/BP-PZN2.0 nanocomposites after exposure to ambient conditions for four months

after exposure to ambient conditions for four month. As a consequence, these characterizations demonstrate that the BP-PZN nanosheets in EP/BP-PZN nanocomposites exhibit good air stability.

2.4 Summary

We have developed a realizable one-pot method to prepare amino-rich cross-linked polyphosphazene functionalized black phosphorus (BP) via polycondensation, and its structure and composition was confirmed by FT-IR, XRD, Raman, XPS and TEM. Then, the obtained polyphosphazene functionalized black phosphorus (BP-PZN) was incorporated into epoxy resin (EP) to fabricate EP/BP-PZN nanocomposites. Addition of 2 wt% BP-PZN into EP results in the distinct increase of the char yield and the reduction of maximum thermal decomposition rate. In addition, the PHRR, THR, SPR and TSR values for EP/BP-PZN nanocomposites were significantly decreased. TG-FTIR analysis indicated that the release of pyrolysis gas including CO during the burning of EP significantly suppressed after adding BP-PZN. A flame-retardant mechanism was proposed on basis of the analyses for gas and condensed phase products from combustion. It is of importance that the EP/BP-PZN nanocomposites present air stability after exposure to ambient conditions for four month, which was confirmed by XRD pattern and Raman spectra of the nanocomposites. The air stability of the BP nanosheets in EP matrix is attributed to surface wrapping by polyphosphazene and embedding in the polymer matrix as dual protection. The development of a simple, low-cost and economical method resulting in large scale production and surface functionalization of BP nanosheets is achieved in the present study, and the preparation and functionalization strategy offers a promising approach for the design of the high performance BP-polymer based nanocomposites.

References

1. Liu H, Du Y, Deng Y, Peide DY (2015) Semiconducting black phosphorus: synthesis, transport properties and electronic applications. Chem Soc Rev 44:2732–2743
2. Wang H, Yang X, Shao W, Chen S, Xie J, Zhang X, Wang J, Xie Y (2015) Ultrathin black phosphorus nanosheets for efficient singlet oxygen generation. J Am Chem Soc 137:11376–11382
3. Qiao J, Kong X, Hu Z-X, Yang F, Ji W (2014) High-mobility transport anisotropy and linear dichroism in few-layer black phosphorus. Nat Commun 5:4475
4. Ren X, Lian P, Xie D, Yang Y, Mei Y, Huang X, Wang Z, Yin X (2017) Properties, preparation and application of black phosphorus/phosphorene for energy storage: a review. J Mater Sci 52:10364–10386
5. Tao W, Zhu X, Yu X, Zeng X, Xiao Q, Zhang X, Ji X, Wang X, Shi J, Zhang H (2017) Black phosphorus nanosheets as a robust delivery platform for cancer theranostics. Adv Mater 29:1603276
6. Walia S, Balendhran S, Ahmed T, Singh M, El-Badawi C, Brennan MD, Weerathunge P, Karim MN, Rahman F, Rassell A (2017) Ambient protection of few-layer black phosphorus via sequestration of reactive oxygen species. Adv Mater 29:1700152
7. Ryder CR, Wood JD, Wells SA, Yang Y, Jariwala D, Marks TJ, Schatz GC, Hersam MC (2016) Covalent functionalization and passivation of exfoliated black phosphorus via aryl diazonium chemistry. Nat Chem 8:597
8. Nilges T, Kersting M, Pfeifer T (2008) A fast low-pressure transport route to large black phosphorus single crystals. J Solid State Chem 181:1707–1711
9. Aldave SH, Yogeesh MN, Zhu W, Kim J, Sonde SS, Nayak AP, Akinwande D (2016) Characterization and sonochemical synthesis of black phosphorus from red phosphorus. 2D Mater 3:014007
10. Kang J, Wood JD, Wells SA, Lee J-H, Liu X, Chen K-S, Hersam MC (2015) Solvent exfoliation of electronic-grade, two-dimensional black phosphorus. ACS Nano 9:3596–3604
11. Brent JR, Savjani N, Lewis EA, Haigh SJ, Lewis DJ, O'Brien P (2014) Production of few-layer phosphorene by liquid exfoliation of black phosphorus. Chem Commun 50:13338–13341
12. Zhao Y, Wang H, Huang H, Xiao Q, Xu Y, Guo Z, Xie H, Shao J, Sun Z, Han W (2016) Surface coordination of black phosphorus for robust air and water stability. Angew Chem Int Ed 55:5003–5007
13. Avsar A, Vera-Marun IJ, Tan JY, Watanabe K, Taniguchi T, Castro Neto AH, Ozyilmaz B (2015) Air-stable transport in graphene-contacted, fully encapsulated ultrathin black phosphorus-based field-effect transistors. Acs Nano 9:4138–4145
14. Mu H, Lin S, Wang Z, Xiao S, Li P, Chen Y, Zhang H, Bao H, Lau SP, Pan C (2015) Black phosphorus–polymer composites for pulsed lasers. Adv Opt Mater 3:1447–1453
15. Zhu L, Xu Y, Yuan W, Xi J, Huang X, Tang X, Zheng S (2006) One-Pot synthesis of poly (cyclotriphosphazene-co-4, 4′-sulfonyldiphenol) nanotubes via an in situ template approach. Adv Mater 18:2997–3000

16. Tian Z, Hess A, Fellin CR, Nulwala H, Allcock HR (2015) Phosphazene high polymers and models with cyclic aliphatic side groups: new structure–property relationships. Macromolecules 48:4301–4311
17. Qiu S, Wang X, Yu B, Feng X, Mu X, Yuen RK, Hu Y (2017) Flame-retardant-wrapped polyphosphazene nanotubes: a novel strategy for enhancing the flame retardancy and smoke toxicity suppression of epoxy resins. J Hazard Mater 325:327–339
18. Fu J, Huang X, Huang Y, Zhang J, Tang X (2009) One-pot noncovalent method to functionalize multi-walled carbon nanotubes using cyclomatrix-type polyphosphazenes. Chem Commun 1049–1051
19. Gu X, Huang X, Wei H, Tang X (2011) Synthesis of novel epoxy-group modified phosphazene-containing nanotube and its reinforcing effect in epoxy resin. Eur Polymer J 47:903–910
20. Qiu S, Ma C, Wang X, Zhou X, Feng X, Yuen RK, Hu Y (2018) Melamine-containing polyphosphazene wrapped ammonium polyphosphate: a novel multifunctional organic-inorganic hybrid flame retardant. J Hazard Mater 344:839–848
21. Wang J, Liu D, Huang H, Yang N, Yu B, Wen M, Wang X, Chu PK, Yu XF (2018) In-plane black phosphorus/dicobalt phosphide heterostructure for efficient electrocatalysis. Angew Chem 130:2630–2634
22. Zhu X, Zhang T, Sun Z, Chen H, Guan J, Chen X, Ji H, Du P, Yang S (2017) Black phosphorus revisited: a missing metal-free elemental photocatalyst for visible light hydrogen evolution. Adv Mater 29:1605776
23. Hanlon D, Backes C, Doherty E, Cucinotta CS, Berner NC, Boland C, Lee K, Harvey A, Lynch P, Gholamvand Z (2015) Liquid exfoliation of solvent-stabilized few-layer black phosphorus for applications beyond electronics. Nat Commun 6:8563
24. Shao L, Sun H, Miao L, Chen X, Han M, Sun J, Liu S, Li L, Cheng F, Chen J (2018) Facile preparation of NH 2-functionalized black phosphorene for the electrocatalytic hydrogen evolution reaction. J Mater Chem A 6:2494–2499
25. Xu Z-L, Lin S, Onofrio N, Zhou L, Shi F, Lu W, Kang K, Zhang Q, Lau SP (2018) Exceptional catalytic effects of black phosphorus quantum dots in shuttling-free lithium sulfur batteries. Nat Commun 9:4164
26. Wang X, Xing W, Feng X, Song L, Hu Y (2017) MoS2/polymer nanocomposites: preparation, properties, and applications. Polym Rev 57:440–466
27. Qiu S, Xing W, Feng X, Yu B, Mu X, Yuen RK, Hu Y (2017) Self-standing cuprous oxide nanoparticles on silica@ polyphosphazene nanospheres: 3D nanostructure for enhancing the flame retardancy and toxic effluents elimination of epoxy resins via synergistic catalytic effect. Chem Eng J 309:802–814
28. Chen X, Wang W, Jiao C (2017) A recycled environmental friendly flame retardant by modifying para-aramid fiber with phosphorus acid for thermoplastic polyurethane elastomer. J Hazard Mater 331:257–264
29. Ferrari AC, Basko DM (2013) Raman spectroscopy as a versatile tool for studying the properties of graphene. Nat Nanotechnol 8:235
30. Vázquez-Santos MB, Geissler E, László K, Rouzaud J-N, Martínez-Alonso A, Tascón JM (2011) Comparative XRD, Raman, and TEM study on graphitization of PBO-derived carbon fibers. J Phys Chem C 116:257–268
31. Cho SY, Allcock HR (2007) Novel highly fluorinated perfluorocyclobutane-based phosphazene polymers for photonic applications. Chem Mater 19:6338–6344
32. Wu Q, Lü J, Qu B (2003) Preparation and characterization of microcapsulated red phosphorus and its flame-retardant mechanism in halogen-free flame retardant polyolefins. Polym Int 52:1326–1331

Chapter 3
Electrochemically Exfoliated Functionalized Black Phosphorene and Its Polyurethane Acrylate Nanocomposites: Synthesis and Applications

3.1 Introduction

The variety of unique properties of graphene are superior to its bulk counterparts, which have inspired great efforts of researchers to look for alternative new two-dimensional (2D) nanomaterials [1]. Black phosphorus (BP) has an orthogonal fold layer, which is composed by P atoms covalently bonded to three neighbors [2]. The mono-layer or few-layer BP nanosheet called "phosphorene", similar to graphene, can be exfoliated through the scotch-tape microcleavage method from bulk BP [3]. The BP is superior to traditional 2D materials including graphene and transition metal dichalcogenides (TMDs). Phosphorene exhibits high charge mobility (up to 1000 cm^2/V s), thickness-tunable bandgap and anisotropic transport [4]. These intriguing properties display potential applications of phosphorene in optoelectronic and electronic field (such as solar energy conversion, photo detectors and transistors) as well as photocatalysis for water splitting [5]. On the other hand, by virtue of possessing the characteristic size effects, favorable mechanical performances and thermostability, the few-layer BP has the potential to be a new candidate of nanofillers for manufacturing the polymer nanocomposites, which is similar to graphene [6]. For instance, BP nanosheets were exfoliated then encapsulated into a poly(vinyl alcohol) (PVA) matrix, these PVA nanocomposites shown significantly enhanced mechanical properties. It is worth noting that the strengthening effect of BP is better than that of graphene reinforcing PVA with same loading [7]. Up to now, there is rare report about using few-layer BP as nanoadditive for polymer nanocompistes, the practical application of polymer/BP nanocomposites is still a great challenge to be overcome [8]. It is importance of the exfoliation of high quality BP in scalable quantity due to the sizable demand in preparation of polymer/BP nanocomposites.

Studies have been reported on the preparation of multilayer BP crystal by high temperature or high pressure transformation of red phosphorus (RP) or white phosphorus [9, 10]. As same as the graphene from graphite, several techniques have been used to exfoliate monolayer phosphorene or few layer BP nanosheets from original

BP crystal. At present, the top-down strategy is widely used in the preparation of single or few-layer BP through various methods of exfoliation procedures, such as micromechanical cleavage, ball milling, plasma thinning treatment [11], and liquid exfoliation in ionic liquids or organic solvents [12]. The micromechanical method for obtaining the exfoliated BP nanosheets, which features high quality with no structural damage and transverse lateral size, met the requirements but lacked scalability. By the means of ball milling, the few-layer BP nanosheets can prepared, which have edge defects of small lateral size and weak crystallinity [13]. In addition, liquid exfoliation provides a good choice for scalable production but the large consumption in time and the structural defects resulted by prolonged sonication could restrict its application in electronic devices [14]. It is necessary to search alternative synthesis approaches that can produce high quality BP in large quantity. Electrochemical exfoliation is a facile, economic, and environmental method, has been employed to prepare 2D nanomaterial containing graphene, MoS_2 and $Bi_2Se_3(Bi_2Te_3)$ [15], especially for graphene, gratifying achievements in terms of high quality and high yields of the exfoliated nanosheets. Pumera et al. reported the electrochemically exfoliated BP by using H_2SO_4 acidic aqueous solution as electrolyte [16]. Zhao et al. provided a scalable electrochemical method to produce 2D few-layer BP nanosheets where the BP crystal used as the cathode [17]. However, maybe it is a challenge to achieve optimized properties for BP based polymer nanocomposites due to the bare surface nature of these electrochemically exfoliated BP, which is difficult to disperse and incompatible with common polymer matrix. As we know, surface functionalization of graphene served a crucial role on improving the dispersion state and comprehensive properties of polymer nanocomposites [18]. The nanocomposites of BP nanosheets and polymer matrix have difficulties with poor dispersion and poor compatibility, which are similar to graphene based nanocomposites. In order to enhance the compatibility and interfacial interaction, it is a good choice to functionalize the layered material via covalent or non-covalent solutions, which can improve the comprehensive properties of polymer nanocomposites ulteriorly. It is highly desirable to develop a method, including simple, low-cost and environmentally friendly, leading to the large-scale production and surface functionalization of BP nanosheets.

UV-curable polyurethane acrylate (PUA) represents a major trend of polyurethane development owing to the rising attention to human health, environmental pollution and security implications [19]. However, the UV-curable PUA coatings are short of the sufficient flame resistance and mechanical strength in practical applications [20]. There is no report that indicates the influence of BP nanosheets on the mechanical property and flame retardancy of PUA nanocomposites. Phytic acid is biocompatible and eco-friendly material, has been utilized as biological base flame retardant for polymers owing to it is composed of six phosphate groups with rich phosphorus element [21]. Whereas the phytic acid and its derivatives have limited flame retardant efficiency when they used alone. In order to developing high performance UV-curable PUA/BP nanocomposites with superior mechanical property and fire resistance, herein, phytic acid was optioned as the surface modifying agent and effective electrolyte to simultaneously prepare high-quality phytic acid modified BP nanosheets by electrochemical method, then resulting phytic acid modified BP was mixed with Co^{2+} solution to fabricate the cobaltous phytate functionalized BP, defined as BP-EC-Exf. The fabrication process of the BP-EC-Exf is shown in Fig. 3.1. It is expected that this functionalized BP nanosheets could effectively enhance the mechanical and flame retardant of PUA films.

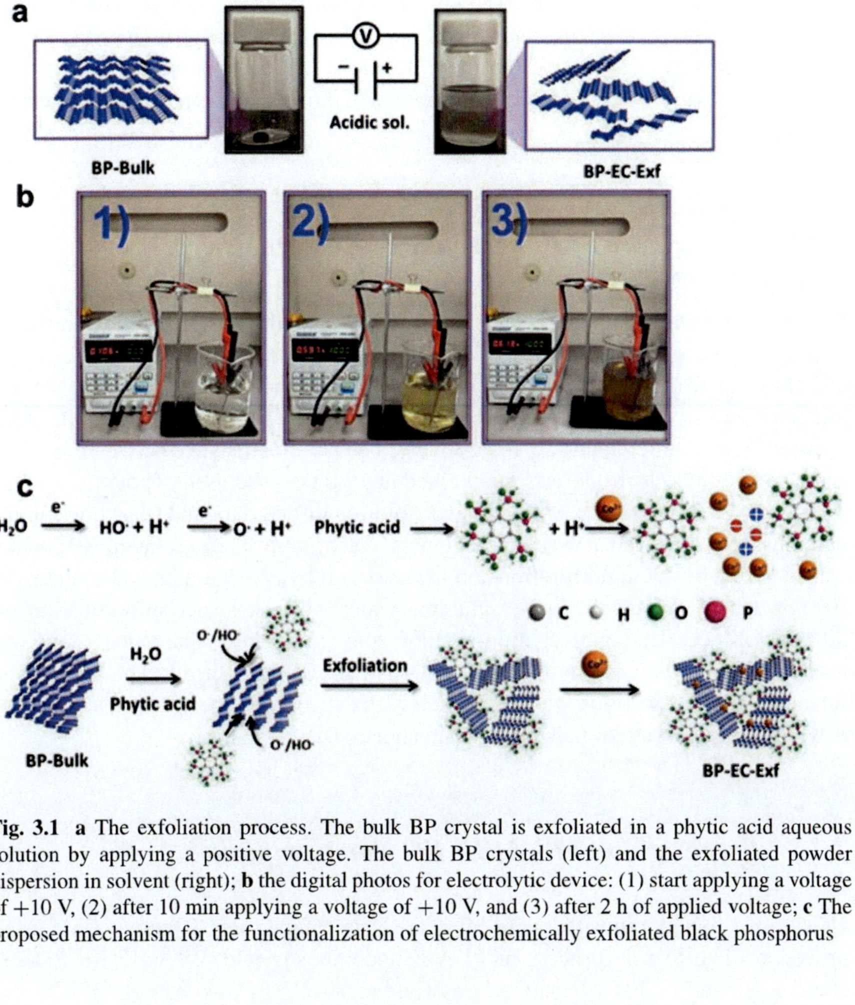

Fig. 3.1 **a** The exfoliation process. The bulk BP crystal is exfoliated in a phytic acid aqueous solution by applying a positive voltage. The bulk BP crystals (left) and the exfoliated powder dispersion in solvent (right); **b** the digital photos for electrolytic device: (1) start applying a voltage of +10 V, (2) after 10 min applying a voltage of +10 V, and (3) after 2 h of applied voltage; **c** The proposed mechanism for the functionalization of electrochemically exfoliated black phosphorus

3.2 Experimental Section

3.2.1 Raw Materials

Phytic acid solution (70%) and $Co(NO_3)_2{\cdot}6H_2O$ were purchased from Aladdin Industrial Corporation (China). PUA resin was purchased from Shanghai Guangyi Chemical co. LTD (China). The photoinitiator, Darocur 1173 was provided by Shanghai Chemical Industry Co., Ltd. (China). Suppliers of other reagents have been described in Chap. 2.

3.2.2 Preparation of Functionalized BP Nanosheets

BP-Bulk was synthesized by a phase transformation method origin from RP utilizing a modified approach as Chap. 2. The BP nanosheets were exfoliated and functionalized by electrochemistry method as follow. The phytic acid solution (0.1 M) was utilized as electrolyte. Large-size BP crystal and copper sheet were served as working electrode and counterelectrode, respectively. The parallel distance between the electrode and counterelectrode was maintained at ~2.0 cm. A continue positive voltage of 10 V was applied for 2 h. After 2 h, the solution turned dark and sparkling minute particles were observed at the electrochemical cell bottom. The exfoliated BP powder was obtained by vacuum filtration and then washed by N_2 saturated water, then, the BP powder was dispersed in N_2 saturated water with assistance of sonication for 30 min followed by centrifugation at 2000 rpm for 10 min. The supernatant was reserved and defined as phytic acid functionalized BP nanosheets. The phytic acid functionalized BP solution was complexed with extra $Co(NO_3)_2$ solution to fabricate cobaltous phytate functionalized BP nanosheets (BP-EC-Exf).

3.2.3 Preparation of PUA/BP-EC Nanocomposites

UV-curable nanocomposites were fabricated via a facile and effective UV curing procedure. Fabrication process of PUA nanocomposite with 0.1 wt% BP-EC-Exf content (named as PUA/BP-EC0.1) as follow: 10 mg of as-prepared BP-EC-Exf powder was dissolved in 20 mL of acetone in a 50 mL three-necked flask under continuous stirring and sonicated (53 kHz) for 1 h under N_2 atmosphere to form a uniform suspension. Then, 9.99 g of PUA was added into the BP-EC-Exf suspension, and the mixture containing PUA and BP-EC-Exf was stirred in ultrasonic bath and maintained for 1 h under N_2 atmosphere. The Darocur 1173 (4 wt%) was dropwise added into the PUA/BP-EC-Exf blends by continuously and vigorously stirring. Subsequently, the residual acetone was removed completely in a vacuum oven at 70 °C for 3 h. Then, the PUA/BP-EC-Exf blends were dip-coated on a glass plate.

Final step is to irradiate the PUA/BP-EC-Exf blends on the glass plate by using an UV irradiation equipment (80 W cm^{-2}, Lantian Co., China). The as-prepared PUA/BP-EC-Exf nanocomposite films were denoted as PUA/BP-ECx, and "x" represents the weight percentage of BP-EC-Exf in PUA nanocomposite films.

3.2.4 Characterization

The combustion properties were evaluated by a microscale combustion calorimeter (MCC) with an instrument model of Govmark MCC-2 (The Govmark Organization, Farmingdale, NY). Approximately 2–5 mg of sample was heated in an inert gas atmosphere at a heating rate of 1 K s^{-1}, then the pyrolysis volatiles with carrier gases (nitrogen of 80 mL min^{-1}; oxygen of 20 mL min^{-1}) were mixed and burned at 900 °C in a combustion chamber. The mechanical property of tensile strength was measured on an MTS CMT6104 universal testing machine (MTS Systems Co. Ltd., P.R. China) according to the Chinese standard of GB 13022−91. The stretching rate was 80 mm min^{-1}. Each specimen was repeated for five times.

Refer to Chap. 2 for FTIR, XRD, Raman, XPS, TEM, SEM, AFM, TGA, DMA and TG-IR measurements.

3.3 Results and Discussion

3.3.1 Morphology and Structure Characterization of BP-EC-Exf

We designed a two electrode equipment where a BP crystal as the anode and copper sheet as the cathode in a 0.1 M phytic acid aqueous solution (Fig. 3.1). There is a parallel distance of 2 cm between two electrodes. An initial voltage of 5 V was applied to the BP crystal for 1 min to promote wetting. With the positive voltage increasing to 10 V, few of minute particles release from the BP crystal slowly and the solution turns yellow (Fig. 3.1b). The electrolysis process tends to stable over time and thus with no need for increasing applied voltage. It is benefit for preventing possible oxidation of BP nanosheets by applying more positive voltage. After 2 h, the color of solution became dark and sparkling minute particles were observed at the electrochemical cell bottom. In this electrochemical exfoliation process, phytic acid was selected as the modifier and electrolyte to prepare functionalized BP. The phytic acid with phosphate groups was able to accelerate the exfoliation procedure and offer the BP nanosheets with flame-retardant function. Subsequently, the resulting phytic acid modified BP was complexed with Co^{2+} ions to fabricate the cobaltous phytate functionalized BP (Fig. 3.1c). The mechanism of electrochemical exfoliation of BP in phytic acid is displayed in Fig. 3.1c. The exfoliation process as follow: Firstly, water was electrolyzed to generate oxygen (O·) and hydroxyl radical (OH·); Secondly, the intercalation of phytic acid and formation of O_2 will lead to significant expansion and exfoliation of BP crystal. Then phytic acid anions intercalated into the interlayer of BP by interacting strongly with edge or surface defects, resulting in the phytic acid modified BP. In terms of further improving the flame retardancy of BP, the intercalated phytic acid was coordinated with Co^{2+} to fabricate cobaltous phytate functionalized BP nanosheets, defined as BP-EC-Exf.

FTIR spectra were utilized to confirm the successful functionalization of BP-EC-Exf. Figure 3.2 shows the FT-IR spectra of BP-Bulk and BP-EC-Exf nanosheets. Except for the absorption peaks of bulk BP, the new representative absorptions at 987 and 1148 cm^{-1} are ascribed to stretching vibration of the P-O and P=O groups in phytic acid, respectively [22]. On the other hand, the two sharp peaks at 987 and 1148 cm^{-1} may correspond to stretching vibration of the P-O and P=O in exfoliated BP with low oxidation degree after electrochemical exfoliation [16].

To meet the demand for identifying the electrochemical exfoliation of BP, Fig. 3.3a, b show comparable XRD patterns and Raman spectra of BP-Bulk and BP-EC-Exf. The orthorhombic phase of BP can be detected in both XRD patterns in Fig. 3.3a. As can be observed from the XRD pattern of BP-Bulk, three intense diffraction peaks and another weak peaks located at 17.0, 34.4, and 52.6°, which are ascribed to (020), (040), and (060) crystal planes of orthorhombic BP [16]. For BP-EC-Exf, its XRD diffraction peaks are broader and weaker, implying that the crystallinity decreases and crystalline size is relative smaller after electrochemical exfoliation. In addition, the (111) plane becomes the strongest diffraction peak rather than (040) plane of the BP-EC-Exf nanosheets, this result reveals that the Van der Waals force between nanolayers within BP-EC-Exf become weaker and decrease the number of layers [23]. Figure 3.3b shows the Raman spectra of the BP-Bulk and BP-EC-Exf. Three characteristic vibrational modes of BP-Bulk: in-plane A_g^2 and B_{2g} vibration modes at 463.9 and 436.8 cm^{-1}, respectively; out-plane A_g^1 mode at

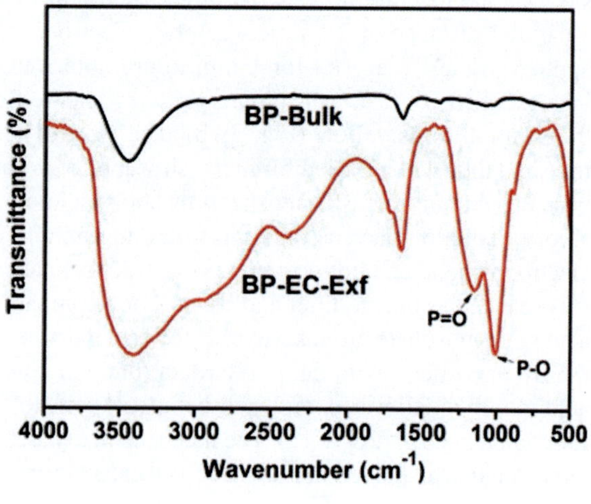

Fig. 3.2 FT-IR spectra of BP-Bulk and BP-EC-Exf nanosheets

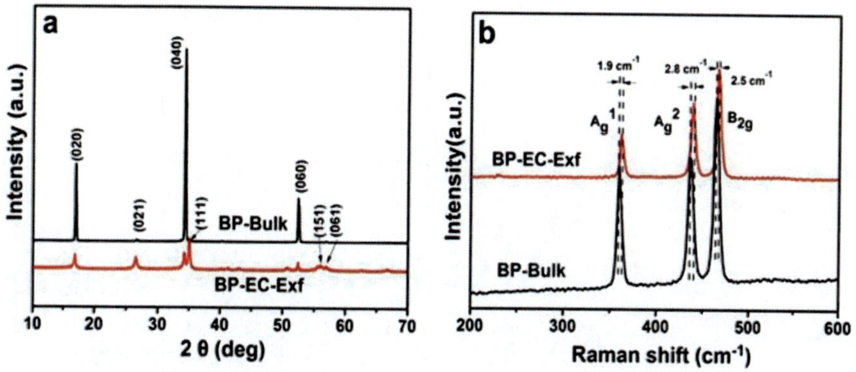

Fig. 3.3 **a** XRD patterns of BP-Bulk and BP-EC-Exf nanosheets and **b** Raman spectra of BP-Bulk and BP-EC-Exf nanosheets

360.1 cm^{-1} [24]. The intensity ratio of A_g^1/A_g^2 is approximately 0.6, this result indicates that the bulk BP exists no oxidation. The blue-shift of the three vibration bands for the BP-EC-Exf demonstrates the decreased layer numbers, which is derived from the minor hindered oscillation of the P atoms owing to decreased interlayer forces [25]. In addition, the intensity ratio of A_g^1 to A_g^2 decreases distinctly for the BP-EC-Exf than the bulk BP, and this phenomenon demonstrates that electrochemically exfoliated BP exhibits a higher oxidation degree or decreased thickness [26].

The mass loss program of BP-EC-Exf was studied by thermogravimetric analysis (TGA) under nitrogen. It shows that bulk BP has one step weight loss between 450 and 550 °C in Fig. 3.4. Nevertheless, BP-EC-Exf has two steps weight loss before 800 °C, the ahead of degradation is observed for BP-EC-Exf, due to the thermal elimination of the phytic acid and other defects on the BP nanosheets derived from electrochemically exfoliation process.

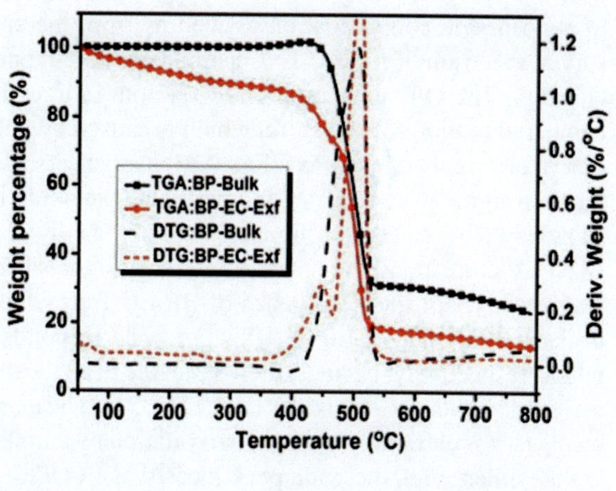

Fig. 3.4 TGA and DTG curves of BP-Bulk and BP-EC-Exf nanosheets

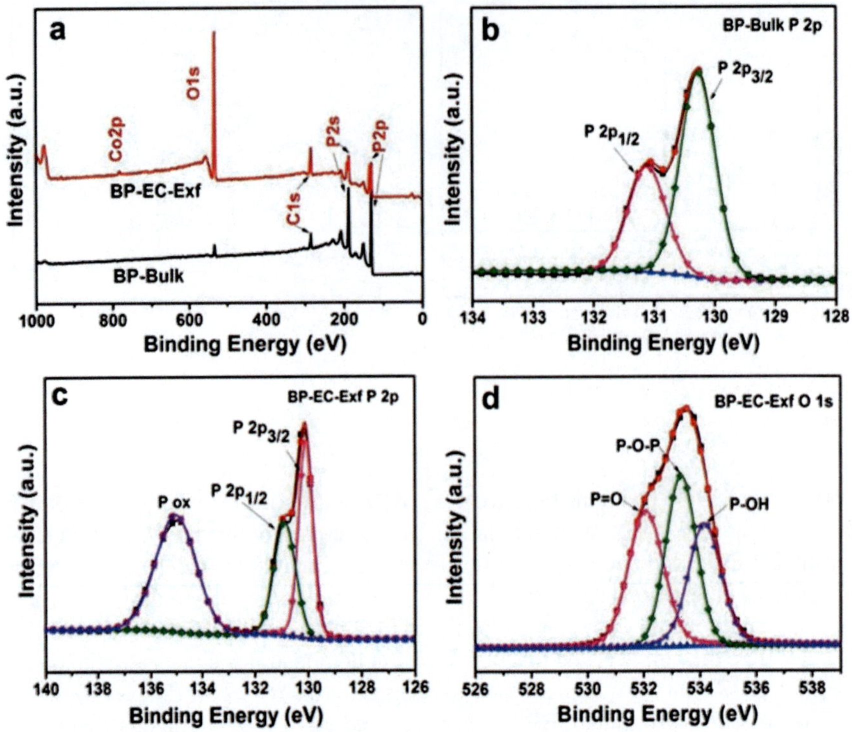

Fig. 3.5 **a** XPS survey spectra of BP-Bulk and BP-EC-Exf; **b**, **c** High-resolution P 2p XPS spectra of BP-Bulk and BP-EC-Exf; **d** High-resolution O 1s XPS spectra of BP-EC-Exf

The BP-Bulk and BP-EC-Exf were studied by XPS to confirm the atomic oxidation state and composition of these materials. XPS survey spectrum of the BP-Bulk only shows the presence of P, C and O in Fig. 3.5a, the existence of C and O is assigned to atmospheric contamination as well as from the specimen holder. However, the survey spectrum of BP-EC-Exf distinctly shows the presence of Co in addition to P, C and O. The O/P atom ratio and C/P atom ratio of BP-EC-Exf sample increased, compared to those of BP-Bulk, which is caused by cobaltous phytate functionalization of electrochemically exfoliated BP nanosheets. High-resolution XPS spectra of the P 2p signal were detected to investigate the oxidation state of P. BP-Bulk crystal shows a well-defined P 2p signal (Fig. 3.5b), are divided into two peaks at 130.2 and 131.1 eV correspond to two binding energy signals P $2p_{3/2}$ and P $2p_{1/2}$ of P-P bonds, respectively. For the P 2p peaks of BP-EC-Exf, which can be divided into several peaks at 130.1 and 131.0 eV (Fig. 3.5c), corresponds to P $2p_{3/2}$ and P $2p_{1/2}$ of P-P bonds, respectively; besides, there is a wide peak located at 134.9 eV, which could be assigned to phosphorus oxides (PO_x) [17, 24]. The existence of PO_x is normal for BP owing to it is extremely inclined to oxidation by ambient species (H_2O/O_2) in the air. On the other hand, the wide peak located at 134.9 eV for BP-EC-Exf sample could be attributed to P-O and P=O bonding in the phytic acid. Fitting analysis reveals that the O 1s peaks of the BP-EC-Exf (Fig. 3.5d) are divided into several peaks at 533.3 and 532.1 eV, which correspond to P-O-P and P=O bonds, respectively. Except for the P-O-P and P=O signals, there exist an extra peak at 534.1 eV corresponds to P-OH bond, due to the hydroxyl groups of phytic acid which exhibits on the surface or interlayer of BP-EC-Exf nanosheets, revealing the successful formation of cobaltous phytate functionalized BP nanosheets.

The SEM image in Fig. 3.6a illustrates the bulk BP crystal has multilayer structure with uneven size. TEM image in Fig. 3.6b presents a typical micromorphology of the electrochemically exfoliated BP nanosheets. It can be observed clearly that lamellar BP nanosheets are obtained after the exfoliation. The BP-EC-Exf sample shows transparent layered sheets for electrons, revealing a few numbers of layers. The high-resolution TEM (HRTEM) image is shown in Fig. 3.6c with a selected area electron diffraction (SAED) pattern in the inset, the corresponding image manifests the orthogonal crystal structure of BP nanosheets, and the lattice stripe of 0.25 nm is distributed to the (111) surface of BP, which is consistent with the XRD analysis results. Figure 3.6d presents the model of layered structure of multilayer BP.

A magnified AFM image and the height profiles of electrochemically exfoliated BP nanosheets are shown in Fig. 3.7a, b. The BP nanosheets express a laminar structure with an average thickness of 6.08 nm. Generally, the thickness of a single layer of

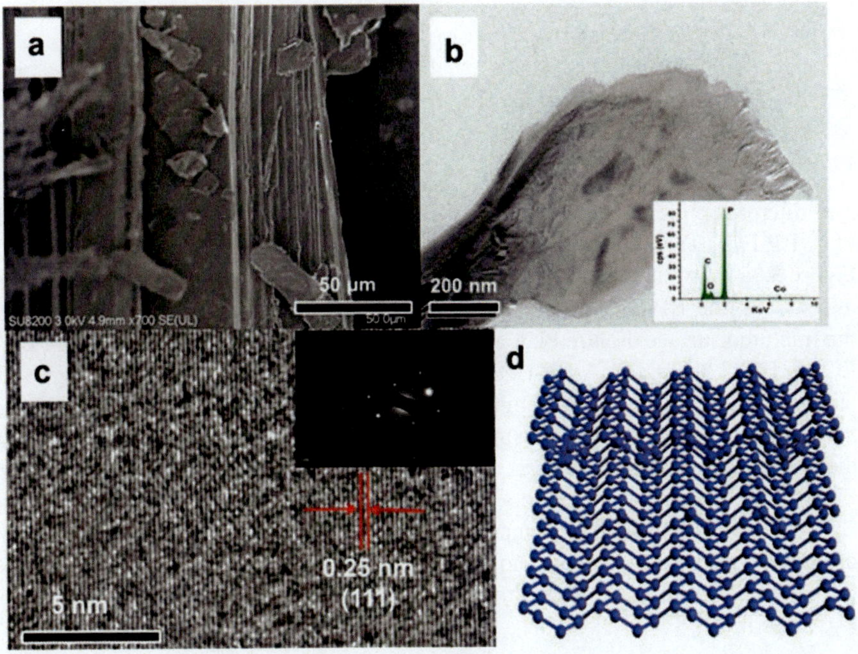

Fig. 3.6 **a** SEM image of the BP crystal; **b** TEM image the BP-EC-Exf nanosheets; **c** HRTEM image and the corresponding SAED pattern of the BP-EC-Exf nanosheets; **d** the structure model of atomic layer for folded-multilayer BP

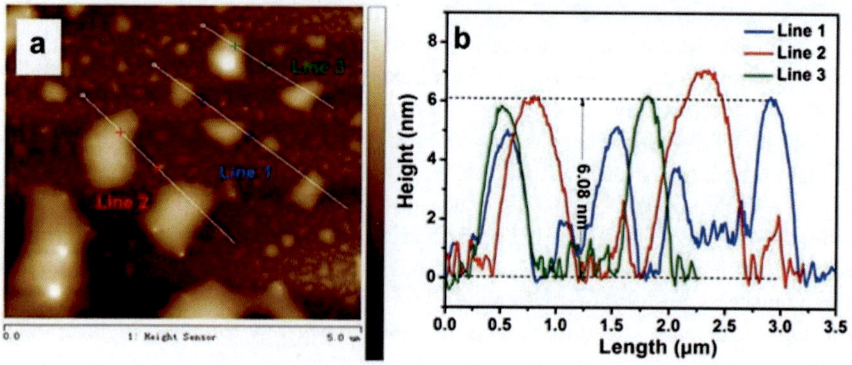

Fig. 3.7 **a** AFM image of BP-EC-Exf nanosheets and **b** height profiles taken along the lines marked in (a)

black phosphorene is thought to be 0.53 nm, corresponds to 11 layers of P atoms [23]. Such exfoliated BP could be defined as few-layer black phosphorene materials. In addition, BP-EC-Exf nanosheets have an averaged size of several hundred nanometers. These credible results indicate that electrochemical exfoliation is a one-pot and high-efficiency method for obtaining few-layer BP nanosheets.

3.3.2 Morphology Characterization of PUA/BP-EC Nanocomposites

The microstructures about freeze-fractured surface for PUA, PUA/BP-EC0.5 and PUA/BP-EC3.0 nanocomposite films were evaluated by SEM, as shown in Fig. 3.8. Figure 3.8a shows that the fractured surface of pure PUA is extremely smooth. After introducing the BP-EC-Exf into EP, it can be seen the large differences between the fractured surface of pure PUA and PUA/BP-EC nanocomposites, both PUA/BP-EC0.5 (Fig. 3.8b) and PUA/BP-EC3.0 (Fig. 3.8c) samples are shown in the rough and large-crinkled morphologies. Significantly, with increasing the BP-EC-Exf contents, the fractured surface of these samples became rougher. This result demonstrates that introducing BP-EC-Exf has a strong influence on the interface characteristic of PUA/BP-EC composites. It can be observed that there are no aggregations of BP-EC-Exf in the PUA/BP-EC nanocomposites, indicating the well dispersed BP-EC-Exf in PUA matrix. Simultaneously, no visible pull-out of the BP-EC-Exf can be seen as well, which reveals the enhanced interfacial adhesion of the BP-EC-Exf within the PUA matrix.

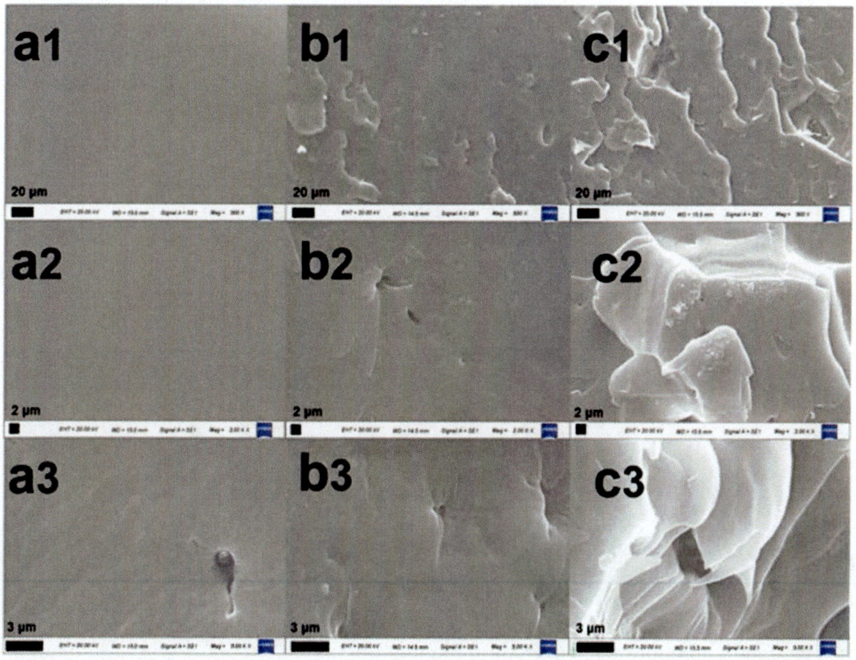

Fig. 3.8 SEM images of the (a1-a3) pure PUA, (b1-b3) PUA/BP-EC0.5 and (c1-c3) PUA/BP-EC3.0 nanocomposites in different magnification

3.3.3 Mechanical Properties of PUA/BP-EC Nanocomposites

Figure 3.9a shows the tensile stress–strain curves of the PUA/BP-EC nanocomposites, and the detail data of tensile strength and tensile fracture strain are listed in Fig. 3.9b, c. The tensile tests demonstrate that the PUA/BP-EC nanocomposites exhibit excellent integration of the mechanical strength. The pure PUA shows a tensile strength of 13.2 MPa, accompany a tensile fracture strain of 42.0%. With increasing BP-EC-Exf content from 0.1 to 3.0 wt%, the tensile strength of the PUA/BP-EC nanocomposites increases from 17.7 to 21.1 MPa; the tensile fracture strain increases from 55.6 to 79.0%. When the nominal content of BP-EC-Exf is 3.0 wt%, the optimal mechanical properties of the PUA/BP-EC nanocomposites are achieved, the corresponding tensile strength and tensile fracture strain increased by 59.8% and 88.1%, respectively. Introduction of BP-EC-Exf into PUA matrix which results in increasing the energy transfer efficiency between BP-EC-Exf nanosheets and PUA.

To know the strengthening effect from 2D BP-EC-Exf nanosheets, a crack extension model is proposed, as shown in 3.10a. When the PUA/BP-EC3.0 nanocomposites suffer from stress, the interface forces between BP-EC-Exf nanosheets and PUA are destroyed firstly, and the BP-EC-Exf nanosheets initiate to slide past each other with increasing tensile force, starting the crack. Meanwhile, the cobaltous phytate part

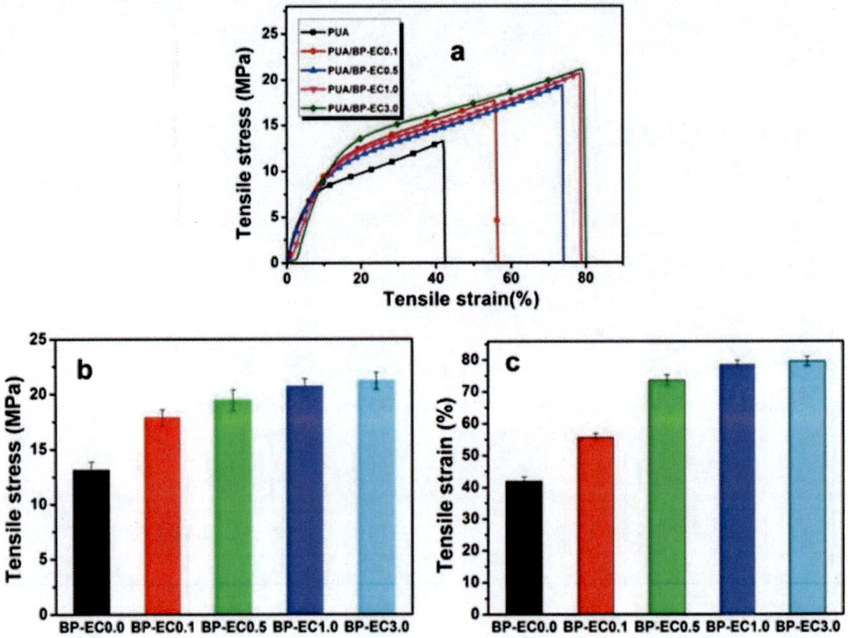

Fig. 3.9 Mechanical properties of the pure PUA and PUA/BP-EC nanocomposites with different BP-EC-Exf contents. **a** Tensile stress–strain curves of the pure PUA and PUA/BP-EC nanocomposites with various BP-EC-Exf contents; **b** Tensile strengths and **c** tensile strains data of the pure PUA and PUA/BP-EC nanocomposites with different BP-EC-Exf contents

with PUA polymer chains can bear excess friction energy dissipation between adjacent BP nanosheets to suppress the sliding effect, and strengthen the PUA/BP-EC3.0 nanocomposites until the cracks forming. Afterwards, the PUA molecular chains are stretched to retard the crack propagation, and to absorb more energy during this stretching procedure. In the next fracture process, 2D BP-EC-Exf nanosheets and PUA polymer chains act a crucial role, which will form crack deflection and lead to the "pulling-out" crack paths (see in Fig. 3.10b). As comparison, the fractured surface of pure PUA after tensile test is extremely smooth with no-crinkled morphology. The digital photos of PUA and PUA/BP-EC films are presented in Fig. 3.10c, the pure PUA film exhibits high transparency whereas the transparency of PUA/BP-EC nanocomposite films are reduced gradually with BP-EC-Exf contents increasing.

DMA was employed to confirm the mechanical and thermal properties were enhanced after incorporating BP-EC-Exf nanosheets into PUA matrix. Figure 3.11 show the storage modulus and tan δ versus temperature curves of the PUA/BP-EC nanocomposites. As shown in Fig. 3.11a, the storage modulus of pure PUA at room temperature is about 1878.7 MPa. The energy storage modulus of PUA/BP-EC nanocomposites are significantly increased by 50.3%, 64.9%, 66.7% and 75.5%, respectively, at room temperature, when the contents of BP-EC-Exf nanosheets increases from 0.1 to 3.0 wt%, compared to neat PUA. The interfacial interaction

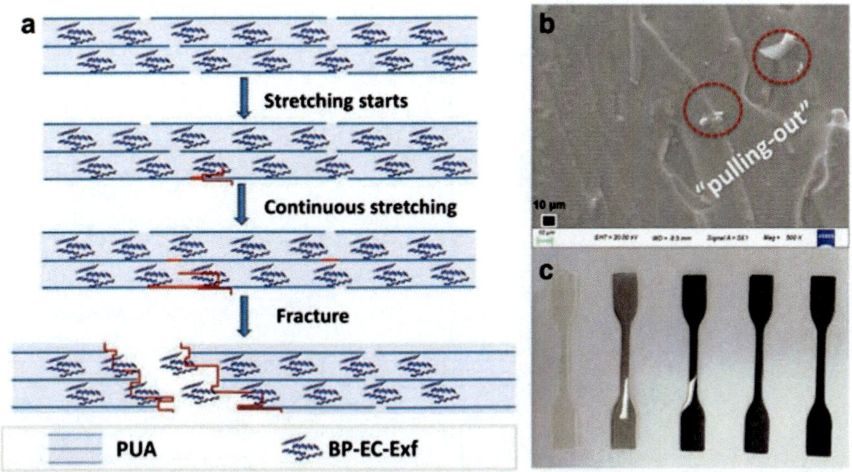

Fig. 3.10 **a** Schematic illustration of the fracture mechanism of the PUA/BP-EC nanocomposites; **b** SEM micrographs of the fracture surface of the PUA/BP-EC3.0 sample after tensile test; **c** the digital photos of pure PUA and PUA/BP-EC nanocomposite films with different BP-EC-Exf contents

between BP-EC-Exf and PUA matrix, and high stiffness of BP-EC-Exf nanosheets are the main reasons to improve the energy storage modulus. As expected, The DMA data for PUA/BP-EC shows higher storage modulus than pure PUA even at higher temperature. The BP-EC-Exf show much better thermal stability due to the well dispersion and inherent characteristics of BP-EC-Exf.

The glass transition temperature (T_g) of PUA/BP-EC nanocomposite can be known from the peak value of tan δ in Fig. 3.11b. In detail, the peak value of tan δ moves slightly to higher temperatures by incorporating the various contents of BP-EC-Exf into PUA matrix. The improvement of the interface interaction between PUA molecular chain and the surface of BP-EC-Exf nanosheet is the main reason for the increase of the T_g values, thereby obtaining the favorable mechanical and thermal performances of PUA/BP-EC composites.

3.3.4 Thermal Stability of PUA/BP-EC Nanocomposites

TGA analysis was employed to confirm the effect of BP-EC-Exf nanosheet on the thermal property of PUA/BP-EC nanocomposites. Upon the Fig. 3.12a, the decomposition behavior of PUA/BP-EC nanocomposites is similar to pure PUA, which has a two-step weight loss process in the area of 300–500 °C. Compared to pure PUA, the initial decomposition temperature of the PUA/BP-EC nanocomposites increased distinctly (see in Fig. 3.12c), with increasing of the BP-EC-Exf contents, the corresponding residual char at 800 °C gradually obviously increased, owing to the catalytic

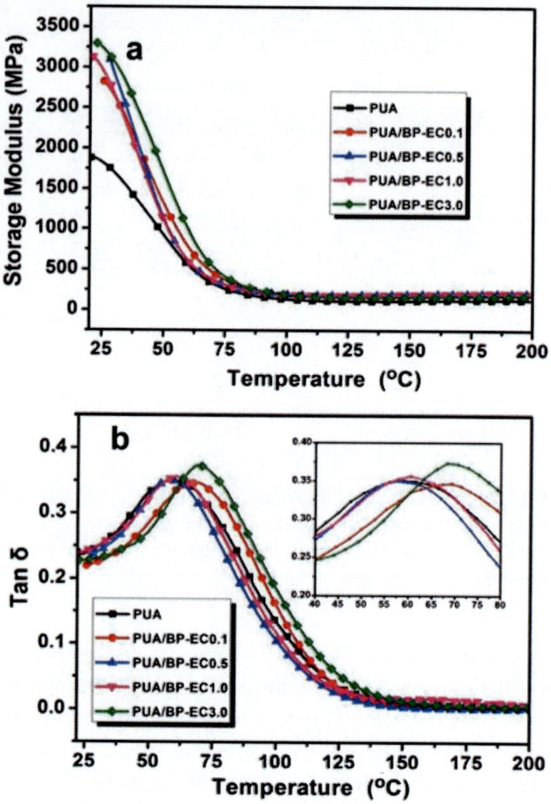

Fig. 3.11 **a** Storage modulus and **b** Tan δ curves of the pure PUA and PUA/BP-EC nanocomposites with different BP-EC-Exf contents as a function of temperature

carbonization effect of cobalt phytic acid and the excellent thermal stability of BP nanoparticles. Besides, the maximum mass loss rates of PUA/BP-EC nanocomposites are shift to lower values, compared to pure PUA, which can be observed in the derivative thermogravimetric analysis (DTG) curves in Fig. 3.12b, d. As a physical barrier during combustion, the exfoliated BP nanosheets can hinder oxygen transport and release of decomposition products, and cobaltous phytate as charring catalyst which promoting the char formation, thus enhancing the thermal stability of the PUA composites against degradation.

3.3.5 *Flame Retardancy of PUA/BP-EC Nanocomposites*

The fire performance of PUA/BP-EC nanocomposites was evaluated by MCC test. Figure 3.13a, b show the heat release rate (HRR) and total heat release (THR) versus temperature curves of PUA nanocomposites. Pure PUA burns dramatically, exhibits high peak heat release rate (PHRR) (355.4 W/g) and total heat release (THR)

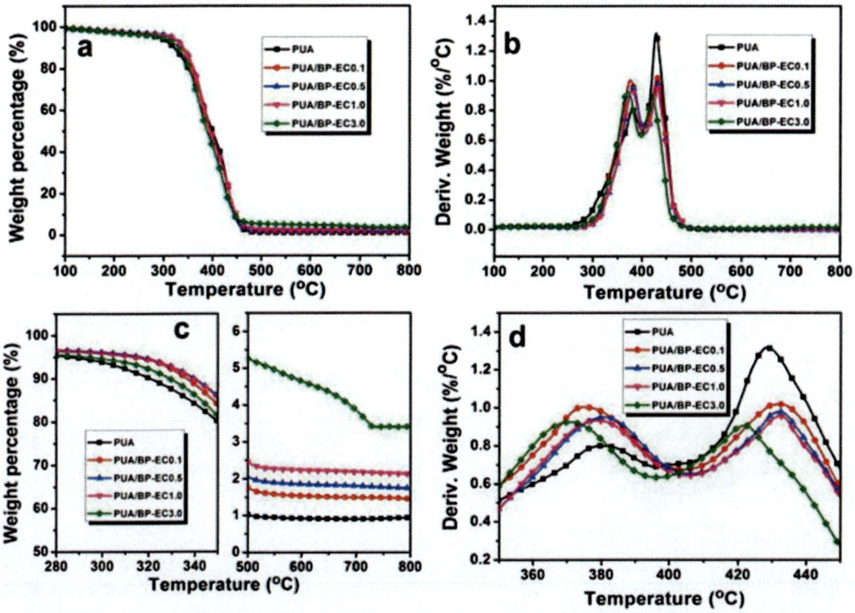

Fig. 3.12 **a** TGA and **b** DTG curves, **c** The TGA curves at 280–350 °C and 500–800 °C regions and **d** the DTG curves at 350–450 °C of the pure PUA and PUA/BP-EC nanocomposites with different BP-EC-Exf contents

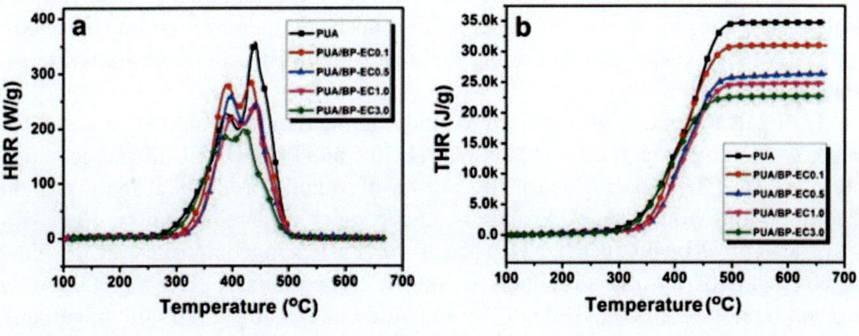

Fig. 3.13 **a** HRR and **b** THR versus temperature curves of the pure PUA and PUA/BP-EC nanocomposites with various BP-EC-Exf contents

(34.9 kJ/g) values, respectively. The PHRR values of the PUA/BP-EC nanocomposites decreased by 20.1–44.5%, with increasing BP-EC-Exf contents from 0.1 to 3.0 wt%; besides, the THR values of the PUA/BP-EC nanocomposites decreased by 11.3 to 34.5%. As expected, the incorporation of 3.0 wt% BP-EC-Exf in PUA, it leads to the maximum reduction in PHRR and THR during combustion, with the increase of BP-EC-Exf content, the flame retardant properties of PUA/BP-EC nanocomposites

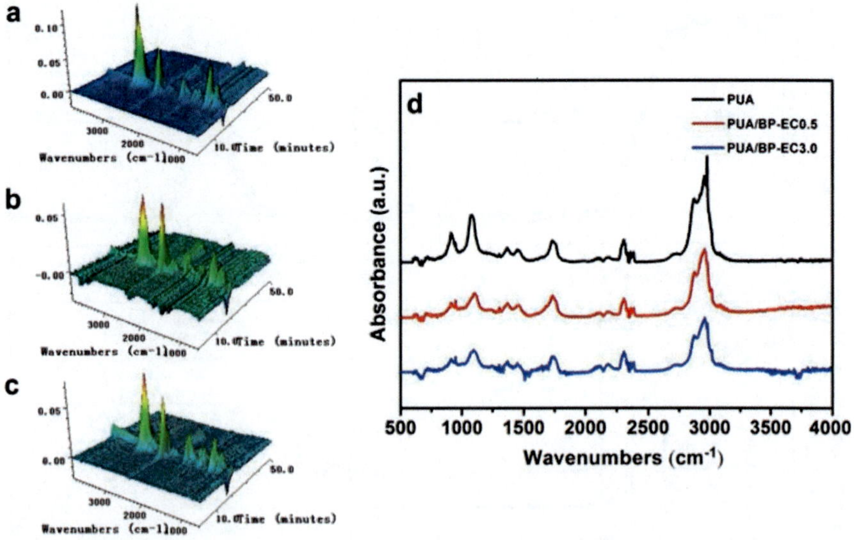

Fig. 3.14 Three-dimensional (3D) TG-FTIR spectra of gasified pyrolysis products for **a** pure PUA, **b** PUA/BP-EC0.5 and **c** PUA/BP-EC3.0. **d** FTIR spectra of gasified pyrolysis products for pure PUA, PUA/BP-EC0.5 and PUA/BP-EC3.0 nanocomposites at the maximum evolution rate

were improved. The notable enhancements in the flame retardancy of PUA nanocomposites are depend on the synergistic catalytic carbonization effect of cobalt phytate with BP and the physical barrier effect of BP nanosheets, thereby retarding the release of pyrolysis products and forbidding the inner unburned materials exposed to fire during combustion.

TG-FTIR technique was utilized to study the thermal decomposition and diffusion of volatile products of PUA, PUA/BP-EC0.5 and PUA/BP-EC3.0 nanocomposites. 3D TG-FTIR spectra and FTIR spectra of volatile products at the maximum evolution rate for pure PUA and its nanocomposites are shown in Fig. 3.14. Several typical peaks in the FTIR spectra of PUA nanocomposites are assigned to characteristic pyrolysis products as follow: carbon dioxide (CO_2) (2360 cm^{-1}); carbon monoxide (CO) (2190 cm^{-1}); aromatic compounds (1510 cm^{-1}); carbonyl compounds (1740 cm^{-1}) and hydrocarbons (2930 cm^{-1}) [27].

The intensities of the typical pyrolysis products versus time curves for PUA nanocomposites are presented in Fig. 3.15. Compared with pure PUA, after adding 0.5 wt% BP-EC-Exf and 3.0 wt% BP-EC-Exf to PUA, the maximum absorbance intensity of characteristic volatile products such as CO, hydrocarbons, carbonyl and aromatic compounds are significantly reduced. The absorption intensity of the above volatile products for PUA/BP-EC3.0 was lowest compared to pure PUA and PUA/BP-EC0.5 nanocomposites. The main toxic substance in the combustion process of PUA is considered as CO. The decrease of CO concentration is beneficial to the reduction of smoke toxicity. At the same time, the release of flammable pyrolysis gaseous

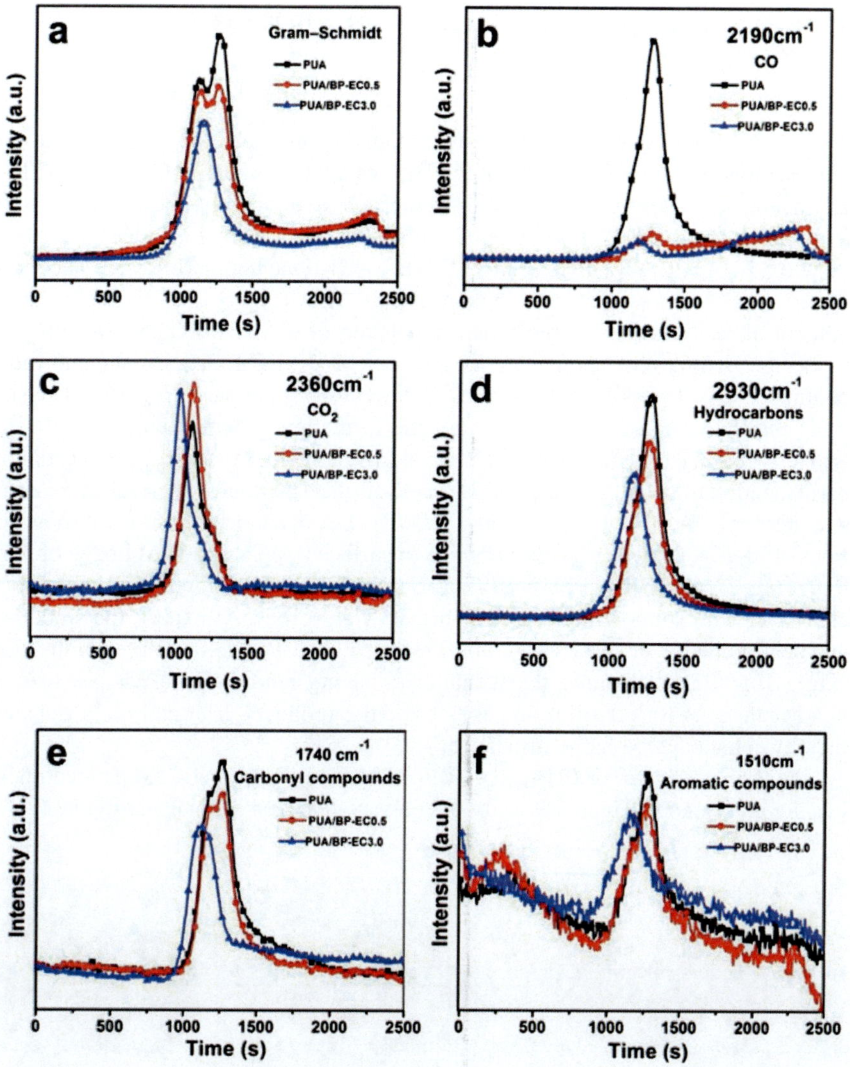

Fig. 3.15 Absorbance of volatile products for pure PUA, PUA/BP-EC0.5 and PUA/BP-EC3.0 nanocomposites sample versus time: **a** total pyrolysis products; **b** CO; **c** CO_2; **d** hydrocarbons; **e** carbonyl compounds and **f** aromatic compounds

products such as hydrocarbons, carbonyls and aromatic compounds is reduced, which contributes to reduce heat release and smoke to improve flame retadancy [28].

3.3.6 *Condensed Phase Analysis of PUA/BP-EC Nanocomposites*

To know the reasonable mechanism of flame retardation, the char residues of PUA nanocomposites were evaluated. The residual chars of PUA and PUA/BP-EC nanocomposites were acquired by thermal treatment of these materials in a muffle furnace at 450 °C. Digital photos of the external residues from top view for PUA, PUA/BP-EC0.5 and PUA/BP-EC1.0 nanocomposites are shown in Fig. 3.16. Few of residual char remain in pure PUA (Fig. 3.16a), however, larger amount of residues are formed after combustion of PUA/BP-EC0.5 nanocomposites (Fig. 3.16b). With increasing the content of BP-EC-Exf, a denser and more continuous char layers are generated in PUA/BP-EC3.0 sample (Fig. 3.16c). Considering the PUA/BP-EC nanocomposites, the formation of high quality char layers is due to BP nanosheets' barrier effect and the effect of the synergistic catalytic carbonization of cobalt phytate and BP nanosheets. The microstructures of external residues for PUA, PUA/BP-EC0.5 and PUA/BP-EC3.0 nanocomposites are shown in Fig. 3.16d–f. A flaky char layer with broken surface is shown in SEM image of pure PUA (Fig. 3.16d). For the PUA/BP-EC0.5 sample, shows a more dense continuous char layer with some wrinkle on the surface (Fig. 3.16e). As expect, the smoother and denser char layer surfaces are observed for PUA/BP-EC3.0 sample combustion (Fig. 3.16f). It is well known that a char layer having a more viscous, denser surface is advantageous for retarding release of volatile products, heat and mass transfer, thereby enhancing the flame resistance.

The Raman spectra of PUA, PUA/BP-EC0.5 and PUA/BP-EC3.0 nanocomposites are presented in Fig. 3.17a–c, two representative peaks which are defined as D

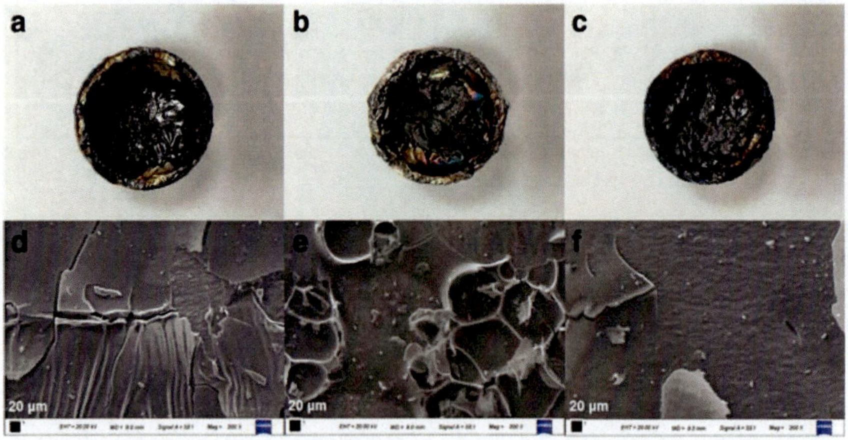

Fig. 3.16 Digital photos of the external char residues from top view for **a** PUA, **b** PUA/BP-EC0.5 and **c** PUA/BP-EC3.0 nanocomposites; SEM images of external residues for **d** PUA, **e** PUA/BP-EC0.5 and **f** PUA/BP-EC3.0 nanocomposites

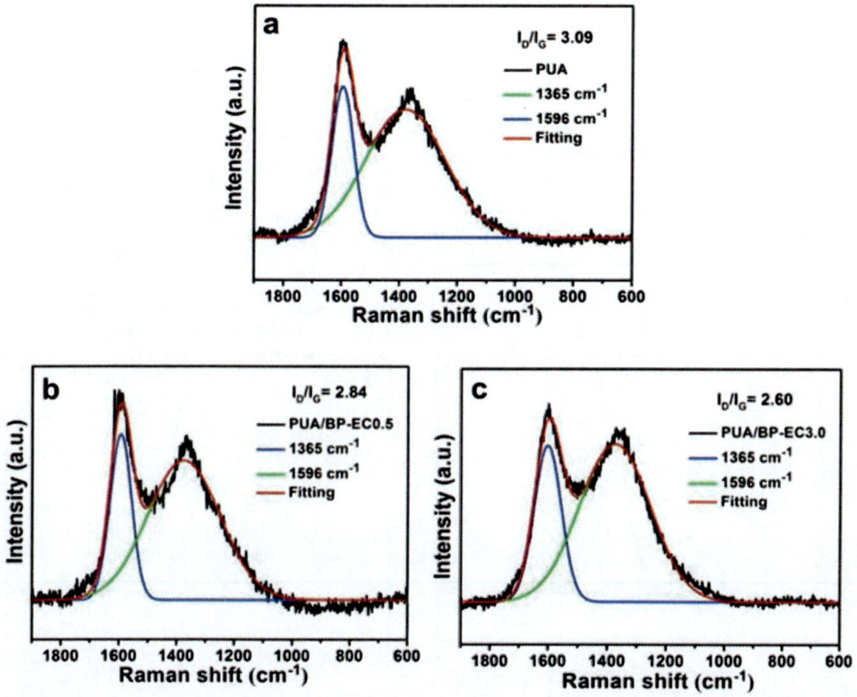

Fig. 3.17 Raman spectra of the residual char for **a** PUA, **b** PUA/BP-EC0.5 and **c** PUA/BP-EC3.0 nanocomposites

and G peak located at 1365 and 1596 cm^{-1}, respectively. In previous studies, the area ratio of D peak to G peak (I_D/I_G) was used to measure the degree of graphitization of residual carbon. The lower the I_D/I_G value, the higher the degree of graphitization [29, 30]. Concretely, the value of I_D/I_G for pure PUA is 3.09, and the PUA/BP-EC0.5 nanocomposite exhibits a I_D/I_G value of 2.84, whereas the lower I_D/I_G value (2.60) of the PUA/BP-EC3.0 nanocomposites in the three samples indicates a higher degree of graphitization. These consequences are ascribed to the catalyzing carbonization effect from both cobaltous phytate and few-layer BP nanosheets during PUA combustion.

Figure 3.18 shows XRD patterns (a) and XPS spectra (b–d) of the residual char for PUA and PUA/BP-EC nanocomposites after combustion in a muffle furnace at 450 °C. The formation of graphitized carbon can be seen from the XRD pattern in which the diffraction peaks of PUA and its nanocomposites are wide, located at about 25.6°, is the graphite based diffraction peaks. Figure 3.18a shows the XRD pattern of PUA, PUA/BP-EC0.5 and PUA/BP-EC3.0, three intense peaks and another weak peaks located at 17.0, 34.1, and 52.4° are observed, corresponds to (020), (040), and (060) crystal planes of orthorhombic BP. It is demonstrate that a part of complete BP nanosheets remain in the char residue of PUA nanocomposites after combustion, which is consistent with the flame-retardant mechanism presented in Fig. 3.19.

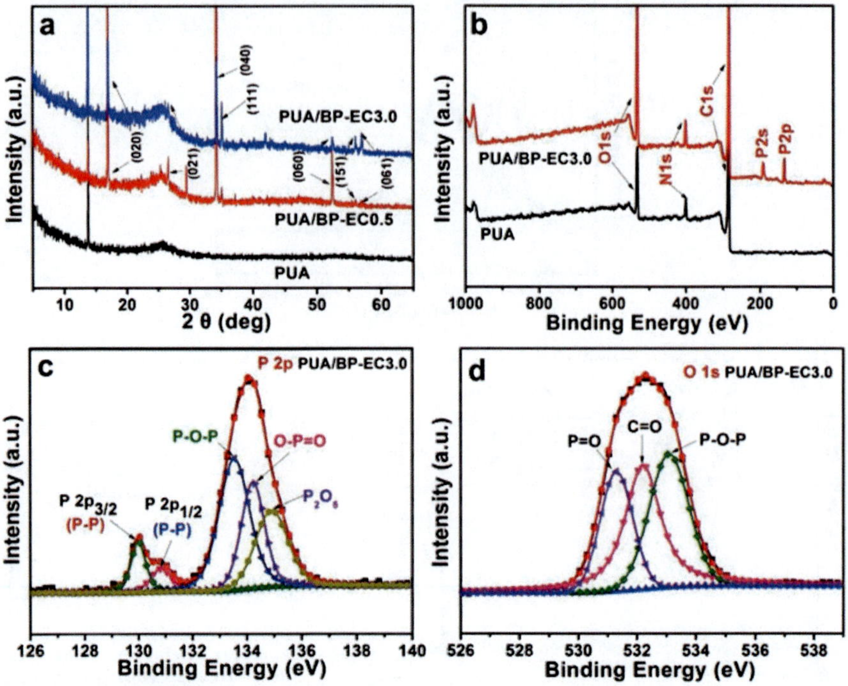

Fig. 3.18 **a** XRD patterns of the char residues for PUA nanocomposites; **b** XPS survey spectra of the residual char for PUA nanocomposites; High-resolution **c** P 2p and **d** O 1s XPS spectra of the residues for PUA/BP-EC3.0 sample

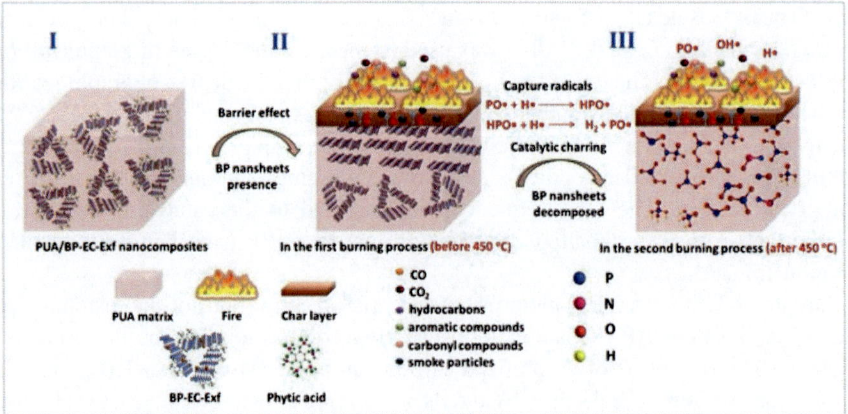

Fig. 3.19 Schematic illustration of flame-retardant mechanism for BP-EC-Exf in PUA nanocomposites during combustion: **I** the PUA/BP-EC-Exf nanocomposites before burning; **II** the first burning process before about 450 °C; **III** the second burning process after about 450 °C of PUA/BP-EC-Exf nanocomposites

More detailed information on the structure and composition of the char residue is provided by XPS analysis. Figure 3.18b shows the XPS spectra of PUA and PUA/BP-EC3.0 nanocomposites, the C, N, O elements can be detected on the surface of the sample, and the extra P element appears in the residue of PUA/BP-EC3.0 nanocomposite, due to the BP and its oxidative derivatives such as PO_x from degradation of BP preserve in the residual char. The high resolution XPS spectrum of the PUA/BP-EC3.0 nanocomposite in the P 2p region is shown in Fig. 3.18c. The P 2p peaks of PUA/BP-EC3.0 are deconvoluted into five peaks at 130.1, 130.9, 133.5, 134.3 and 135.0 eV, corresponds to P $2p_{3/2}$ and P $2p_{1/2}$ of P-P bonds, P-O-P, O-P=O and P_2O_5, respectively, revealing the BP and its oxidative derivatives such as PO_x from the degradation of BP maintain in the residual char. The O 1s XPS spectrum of PUA/BP-EC3.0 in Fig. 3.18d also can verify the above results. There are three strong peaks are ascribed to P=O, C=O and P-O-P bonds, respectively.

3.3.7 Flame Retardant Mechanism of PUA/BP-EC Nanocomposites

A possible flame-retardant mechanism is proposed based on above analysis for thermal decomposition and combustion behavior of PUA/BP-EC nanocomposites (Fig. 3.19). Take the maximum decomposition temperature of BP in air (~450 °C) as the watershed, we divide the combustion process of PUA nanocomposites into two main stages. In the first burning stage before ~450 °C, most of BP nanosheets are remain in the burning sample, BP-EC-Exf nanosheets in PUA matrix act as physical barrier to retard the combustible gases containing aromatic and hydrocarbon fragments release from decomposed PUA nanocomposites to the combustion zone, inhibiting continuous flaming with heat generation. In the second burning stage after ~450 °C, the BP-EC-Exf nanosheets start to decompose gradually. Similar to red phosphorus, the BP has been primarily oxidized into various PO_x and phosphoric acid derivatives with the increase of thermo-oxidative degradation time. The epoxy molecules could react with these phosphoric acid derivatives to generate more stable structures including P—O—C and P—O—P complexes [31]. On the other hand, during the combustion, cobaltous phytate acts as an effective Co-P species catalyst toward the redox reaction in the char layer, which can be confirmed by the significantly decreasing of pyrolysis products, including CO and hydrocarbons. The flame-retardant mechanism demonstrates that the cooperative effect from both cobaltous phytate and few-layer BP nanosheets is the leading cause of the outstanding flame retardancy of PUA nanocomposites.

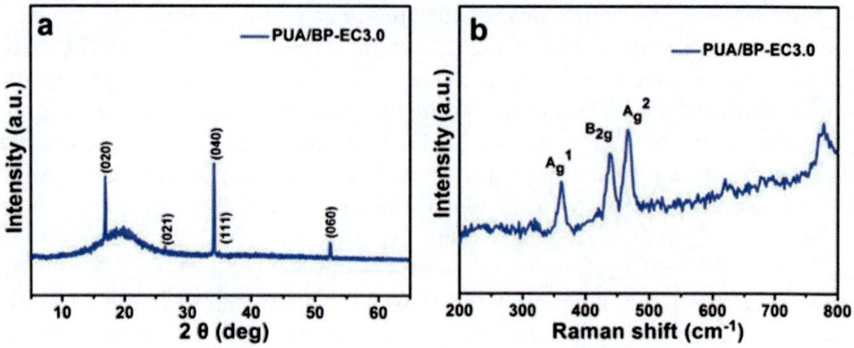

Fig. 3.20 The corresponding **a** XRD pattern and **b** Raman spectrum of PUA/BP-EC3.0 nanocomposites after exposure to ambient conditions for four months

3.3.8 Air Stability of PUA/BP-EC Nanocomposites

To evaluate the air stability of PUA/BP-EC nanocomposites, the PUA/BP-EC3.0 nanocomposites was exposed to atmospheric conditions for four months. The PUA/BP-EC3.0 sample was further investigated by XRD pattern (Fig. 3.20a) and Raman spectrum (Fig. 3.20b). Its XRD pattern exhibits three intense diffraction peaks at 16.9, 34.3 and 52.4° corresponding to the (020), (040), and (060) planes, respectively, which are representative layered planes of BP. In the Raman spectrum, three characteristic vibrational modes of BP correspond to the peaks of A_g^1 at 361.0 cm^{-1}, B_{2g} at 437.3 cm^{-1}, and A_g^2 at 465.7 cm^{-1}. These results confirm that the BP-EC-Exf in PUA/BP-EC nanocomposites with complete structure and crystallinity and cannot be degraded even on prolonged exposure to the air more than four month. As a consequence, these characterizations demonstrate that the BP-EC-Exf nanosheets in PUA/BP-EC nanocomposites exhibit good air stability.

Figure 3.21 shows the SEM image of the film surface **a** and the fractured surface **b** of PUA/BP-EC3.0 sample and their corresponding elemental mapping images after exposure to ambient conditions for four months. The uniformly distributed crystalline BP-EC-Exf nanosheets in the PUA/BP-EC3.0 nanocomposites can be clearly seen by SEM images and elemental mapping images as shown in Fig. 3.21. This characterization demonstrates that the BP-EC-Exf nanosheets in PUA/BP-EC nanocomposites exhibit good air stability.

3.4 Summary

In this work, we developed a facile approach for the electrochemical fabrication of cobaltous phytate functionalized BP nanosheets (BP-EC-Exf) by using BP as the cathode, and phytic acid as surface modifier and electrolyte, simultaneously,

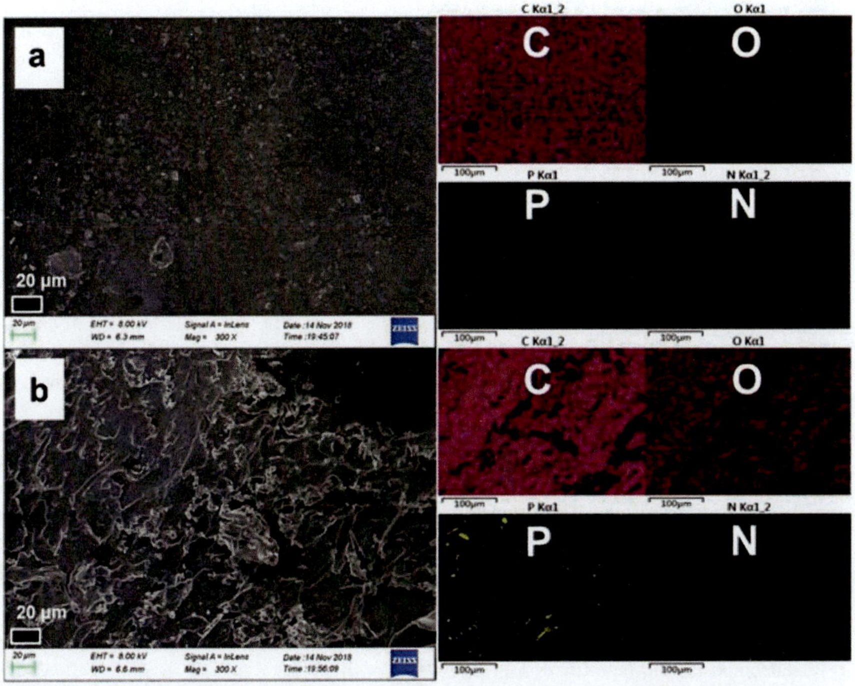

Fig. 3.21 SEM image of the film surface **a** and the fractured surface **b** of PUA/BP-EC3.0 nanocomposites and corresponding elemental mapping images of carbon (C), oxygen (O), phosphorus (P) and nitrogen (N) after exposure to ambient conditions for four months

and their composition and structure were identified by TEM, FT-IR, XRD, XPS and Raman. This resulting BP-EC-Exf provided a multifunctional effect on enhancing the mechanical properties and flame retardancy of PUA nanocomposites. Significantly, conclusion of introducing BP-EC-Exf into PUA matrix resulted in enhancements in mechanical properties of PUA in terms of the tensile strength (increased by 59.8%) and tensile fracture strain (increased by 88.1%); the distinct improvements on flame retardant of PUA in terms of decrease PHRR (reduced by 44.5%) and THR (decreased by 34.5%). TG-FTIR results demonstrated that the release of pyrolysis gas including CO was significantly reduced during combustion after the introduction of BP-EC-Exf. Based on the analysis of gas and condensed phase, a possible flame retardant mechanism was proposed. These significant improvements on fire hazard of PUA origin from the bilateral cooperative effect (physical barrier effect of BP nanosheets and catalytic carbonization action of cobaltous phytate system) of the BP-EC-Exf. It is of importance that the PUA/BP-EC3.0 nanocomposite maintains air stability in environmental condition for four months. Generally, the air stability of BP-EC-Exf nanosheets in PUA can be assigned to surface coating and embedding of BP in PUA matrix as isolation and protection. The functionalization strategy in this study provides a new method to fabricate BP based polymer composites for various applications.

References

1. Englert JM, Dotzer C, Yang G, Schmid M, Papp C, Gottfried JM, Steinrück H-P, Spiecker E, Hauke F, Hirsch A (2011) Covalent bulk functionalization of graphene. Nat Chem 3:279
2. Gusmao R, Sofer Z, Pumera M (2017) Black phosphorus rediscovered: from bulk material to monolayers. Angew Chem Int Ed 56:8052–8072
3. Pang J, Bachmatiuk A, Yin Y, Trzebicka B, Zhao L, Fu L, Mendes RG, Gemming T, Liu Z, Rummeli MH (2018) Applications of phosphorene and black phosphorus in energy conversion and storage devices. Adv Energ Mater 8:1702093
4. Jia J, Jang SK, Lai S, Xu J, Choi YJ, Park J-H, Lee S (2015) Plasma-treated thickness-controlled two-dimensional black phosphorus and its electronic transport properties. ACS Nano 9:8729–8736
5. Xia F, Wang H, Jia Y (2014) Rediscovering black phosphorus as an anisotropic layered material for optoelectronics and electronics. Nat Commun 5:4458
6. Butler SZ, Hollen SM, Cao L, Cui Y, Gupta JA, Gutiérrez HR, Heinz TF, Hong SS, Huang J, Ismach AF (2013) Progress, challenges, and opportunities in two-dimensional materials beyond graphene. ACS Nano 7:2898–2926
7. Ni H, Liu X, Cheng Q (2018) A new strategy for air-stable black phosphorus reinforced PVA nanocomposites. J Mater Chem A 6:7142–7147
8. Ren X, Mei Y, Lian P, Xie D, Yang Y, Wang Y, Wang Z (2018) A novel application of phosphorene as a flame retardant. Polymers 10:227
9. Nilges T, Kersting M, Pfeifer T (2008) A fast low-pressure transport route to large black phosphorus single crystals. J Solid State Chem 181:1707–1711
10. Aldave SH, Yogeesh MN, Zhu W, Kim J, Sonde SS, Nayak AP, Akinwande D (2016) Characterization and sonochemical synthesis of black phosphorus from red phosphorus. 2D Mater 3:014007
11. Hanlon D, Backes C, Doherty E, Cucinotta CS, Berner NC, Boland C, Lee K, Harvey A, Lynch P, Gholamvand Z (2015) Liquid exfoliation of solvent-stabilized few-layer black phosphorus for applications beyond electronics. Nat Commun 6:8563
12. Zhao W, Xue Z, Wang J, Jiang J, Zhao X, Mu T (2015) Large-scale, highly efficient, and green liquid-exfoliation of black phosphorus in ionic liquids. ACS Appl Mater Interfaces 7:27608–27612
13. Sun J, Zheng G, Lee H-W, Liu N, Wang H, Yao H, Yang W, Cui Y (2014) Formation of stable phosphorus–carbon bond for enhanced performance in black phosphorus nanoparticle–graphite composite battery anodes. Nano Lett 14:4573–4580
14. Niu L, Coleman JN, Zhang H, Shin H, Chhowalla M, Zheng Z (2016) Production of two-dimensional nanomaterials via liquid-based direct exfoliation. Small 12:272–293
15. Ambrosi A, Sofer ZK, Luxa J, Pumera M (2016) Exfoliation of layered topological insulators Bi_2Se_3 and Bi_2Te_3 via electrochemistry. ACS Nano 10:11442–11448
16. Ambrosi A, Sofer Z, Pumera M (2017) Electrochemical exfoliation of layered black phosphorus into phosphorene. Angew Chem Int Ed 56:10443–10445
17. Xiao H, Zhao M, Zhang J, Ma X, Zhang J, Hu T, Tang T, Jia J, Wu H (2018) Electrochemical cathode exfoliation of bulky black phosphorus into few-layer phosphorene nanosheets. Electrochem Commun 89:10–13
18. Kang SM, Park S, Kim D, Park SY, Ruoff RS, Lee H (2011) Simultaneous reduction and surface functionalization of graphene oxide by mussel-inspired chemistry. Adv Func Mater 21:108–112

19. Choi S-J, Kim HN, Bae WG, Suh K-Y (2011) Modulus-and surface energy-tunable ultraviolet-curable polyurethane acrylate: properties and applications. J Mater Chem 21:14325–14335
20. Alongi J, Di Blasio A, Carosio F, Malucelli G (2014) UV-cured hybrid organic–inorganic layer by layer assemblies: effect on the flame retardancy of polycarbonate films. Polym Degrad Stabil 107:74–81
21. Guo Q, Cao J, Han Y, Tang Y, Zhang X, Lu C (2017) Biological phytic acid as a multifunctional curing agent for elastomers: towards skin-touchable and flame retardant electronic sensors. Green Chem 19:3418–3427
22. Song X, Chen Y, Rong M, Xie Z, Zhao T, Wang Y, Chen X, Wolfbeis OS (2016) A phytic acid induced super-amphiphilic multifunctional 3D graphene-based foam. Angew Chem Int Ed 55:3936–3941
23. Shao L, Sun H, Miao L, Chen X, Han M, Sun J, Liu S, Li L, Cheng F, Chen J (2018) Facile preparation of NH_2-functionalized black phosphorene for the electrocatalytic hydrogen evolution reaction. J Mater Chem A 6:2494–2499
24. Zhu X, Zhang T, Sun Z, Chen H, Guan J, Chen X, Ji H, Du P, Yang S (2017) Black phosphorus revisited: a missing metal-free elemental photocatalyst for visible light hydrogen evolution. Adv Mater 29:1605776
25. Favron A, Gaufrès E, Fossard F, Phaneuf-L'Heureux A-L, Tang NY, Lévesque PL, Loiseau A, Leonelli R, Francoeur S, Martel R (2015) Photooxidation and quantum confinement effects in exfoliated black phosphorus. Nat Mater 14:826
26. Lu W, Nan H, Hong J, Chen Y, Zhu C, Liang Z, Ma X, Ni Z, Jin C, Zhang Z (2014) Plasma-assisted fabrication of monolayer phosphorene and its Raman characterization. Nano Res 7:853–859
27. Qiu S, Wang X, Yu B, Feng X, Mu X, Yuen RK, Hu Y (2017) Flame-retardant-wrapped polyphosphazene nanotubes: a novel strategy for enhancing the flame retardancy and smoke toxicity suppression of epoxy resins. J Hazard Mater 325:327–339
28. Qiu S, Xing W, Feng X, Yu B, Mu X, Yuen RK, Hu Y (2017) Self-standing cuprous oxide nanoparticles on silica@polyphosphazene nanospheres: 3D nanostructure for enhancing the flame retardancy and toxic effluents elimination of epoxy resins via synergistic catalytic effect. Chem Eng J 309:802–814
29. Ferrari AC, Basko DM (2013) Raman spectroscopy as a versatile tool for studying the properties of graphene. Nat Nanotechnol 8:235
30. Vázquez-Santos MB, Geissler E, László K, Rouzaud J-N, Martínez-Alonso A, Tascón JM (2011) Comparative XRD, Raman, and TEM study on graphitization of PBO-derived carbon fibers. J Phys Chem C 116:257–268
31. Wu Q, Lü J, Qu B (2003) Preparation and characterization of microcapsulated red phosphorus and its flame-retardant mechanism in halogen-free flame retardant polyolefins. Polym Int 52:1326–1331

Chapter 4
Strengthening of Black Phosphorus/Nanofibrillar Cellulose Composite Film with Nacre-Inspired Structure and Superior Fire Resistance

4.1 Introduction

Nacre is an organic–inorganic natural layer structure material, which created from the long-term evolution of nature. The comprehensive toughness and strength of natural nacre are ascribed to its multi-stage ordered "brick-and-mortar" assembly structure, which contains 5 vol% biological polymers including nanofibrillar chitin and proteins, and 5 vol% inorganic aragonite platelets [1, 2]. Thus, nacre, the crucial standard for bionics, offers an optimized guidelines and templates for the assembly of two dimensional (2D) nanosheets into nanocomposites with light weight, high strength, good toughness and excellent mechanical properties [3]. Inspired by the nacre with "brick-and-mortar" layered structure, the researchers prepared a series of bionic high-strength and super-toughness layered composites by different methods [4, 5].

As a promising 2D nanomaterial similar to graphene, few-layer black phosphorus (BP) can be exfoliated from bulk BP, has attracted increasing interest recently owing to its distinct properties, including large on/off ratios, tunable electronic bandgap, anisotropic responses, and high carrier mobility [6]. The lamellar BP nanosheets, defined as "phosphorene", rendering their broad application prospects in energy conversion, storage, biomedicine and transistors [7, 8]. The BP is superior to conventional 2D materials containing transition metal dichalcogenides and graphene, in the field of electrical and optical devices. For the practical application, few-layer 2D material is usually superior to its bulk counterparts owing to its richer exposed active sites and larger specific surface area [9]. Since the bulk BP consists of folded monolayer integrated by weak van der Waals forces, lamellar BP nanosheets have been exfoliated from BP crystal by several methods, including liquid exfoliation, mechanical cleavage, plasma etching and photochemical [10], which enabled specific morphology and layer numbers to be achieved, but are usually time-consuming. Moreover, residual organic pollution often maintain on the surface of few-layer BP [11]. There is no reactive groups on the surface of BP nanosheets, for example, hydroxyl group, amino group (NH_2) and epoxy group, et al., so it is difficult to further modified their surface by reacting with organic compounds or polymer molecular. Similar to traditional 2D layered materials like graphene, the BP nanosheets has poor dispersion and compatibility in organic solvents or polymers, which restrict to fabricate polymer nanocomposites [12]. Ball milling method is another practical strategy for preparing and peeling 2D layered nanomaterials. Herein, Cheng et al. have reported a functionalized BP nanosheets via urea-assisted ball-milling strategy and investigated the electrocatalytic HER properties of BP nanosheets in alkaline media [13]. Meanwhile, the exfoliated BP is functionalized with amino (NH_2) groups at the surface or edge of BP, inhibiting its restacking and forming ultrathin black phosphorene. A recent work demonstrated the LiOH-assisted ball-milling synthesis of BP nanosheets, the exfoliated BP is functionalized with hydroxyl groups at the edge of BP [14]. Refer to these functionalized methods of BP, it is significant to develop facile, economical and green strategies to obtain BP-based functional materials for nanocomposite applications.

Cellulose is the renewable and most abundant natural material which widely distribute on earth [15, 16]. It has many advantages of rich resources, as well as no-toxic, biodegradable, green and environmental protection. Nanofibrillar cellulose (NFC), is derived from cellulose, which can be utilized as a reinforcing material for preparing nanocomposites due to its high flexibility and mechanical strength [17, 18]. Nanofibrillar cellulose, as one-dimensional (1D) building blocks, could be integrated with 2D materials, such as reduced graphene oxide (rGO), graphene and boron nitride (BN) to build binary or ternary artificial nacre through self-assembling strategy, which acts a crucial role in preparing high-performance nanocomposites [19, 20]. For example, a nacre-inspired structure consist of the ultrathin $Ti_3C_2T_x$ (d-$Ti_3C_2T_x$, MXene) and cellulose nanofiber (CNF) was prepared via self-assembly menthd with a vacuum-assisted filtration procedure, the resulting composite film possess high tensile strength and tensile fracture strain, can reach 135.4 MPa and 16.7%, respectively, thus the synergistic strengthening and toughening effect of the composite film has been achieved [21]. An artificial nacre-like nanocomposite was designed through binary strengthening and toughening of 1D cellulose nanofiber and 2D graphene oxide nanosheet as cross-linking building blocks. This resulting bioinspired nanocomposite exhibits superior mechanical properties with high tensile strength and high toughness [22]. However, to our knowledge, there were no immediate reports on the BP/NFC composite film for reinforcing materials and fire insulation application.

The binary artificial nacre was built by assembly of 1D and 2D building blocks with hydroxyl bonding, it plays an important role in the preparation of high performance nanocomposites. This study takes NFC as a one-dimensional building block. NFC is renewable, environmentally friendly and easy to prepare, mainly made of nanometer or micro-scale cellulose, and its elastic modulus up to ~150 GPa [23]. Hydroxyl functionalized BP (BP-OH) nanosheets are benificial to constructing interface interactions with NFC with abundant hydroxyl groups via hydrogen bonding network. A nacre-like bioinspired nanocomposite was designed via synergistic strengthening of 2D BP-OH nanosheets and 1D NFC as building blocks and hydrogen bonding network. The fabrication process of the BP-OHx/NFC ($x = 0, 5, 10, 25, 40$) composite

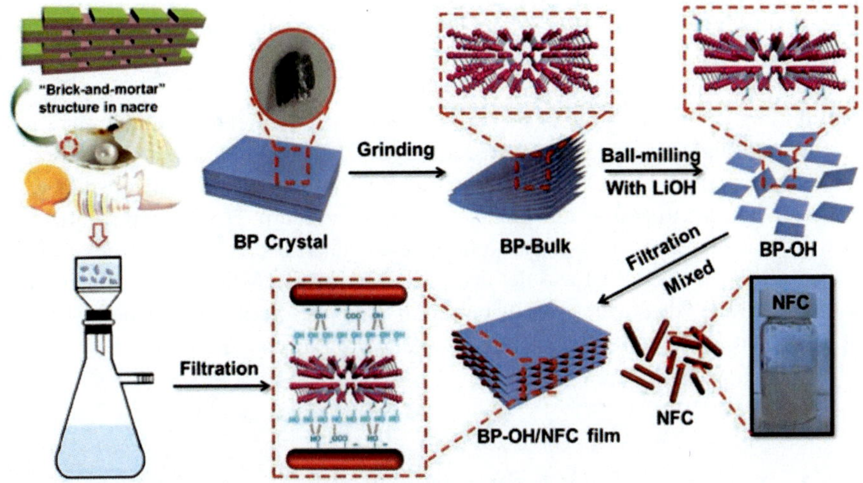

Fig. 4.1 Illustration of the fabrication process of the BP-OHx/NFC composite film

films is shown in Fig. 4.1. These novel binary artificial nacre nanocomposites show the maximum stress of 214.0 MPa and a tensile fracture strain of 23.8%. Meanwhile, these composite films also demonstrate high thermal stability and good fire retardancy. The novel high strength and fire resistance artificial nacre supports have promising potential in diverse application fields including stretchable electronics, biomaterials and flame retardant insulation.

4.2 Experimental Section

4.2.1 Raw Materials

$LiOH{\bullet}H_2O$ were purchased by Aladdin Industrial Corporation (China), $LiOH{\bullet}H_2O$ was subjected to vacuum heating at 110 °C for 4 h to remove the crystallized water. Suppliers of other reagents have been described in Chap. 2.

4.2.2 Preparation of BP-Bulk and BP-OH Nanosheets

BP-Bulk was synthesized by a phase transformation reaction origin from RP utilizing a modified approach in Chap. 2. BP-OH was synthesized by ball-milling solid phase mechanochemical method. In detail, 840 mg LiOH powder with no crystallized water and 360 mg BP-Bulk powder was put and mix in an agate ball-milling jar containing agate balls (various sizes). The agate ball-milling jar was sealed in a glovebox filled fully with argon, and then equipped on the planetary ball-milling machine. The rotational speed set of 300 rpm and the whole ball-milling process was performed at ambient temperature for total 24 h, grind every 30 min, 15 min apart. After the ball-milling is complete, the agate jar was cooled to sufficient temperature at room temperature and then carefully opened in the air. Then, N_2 saturated deionized water was poured in quickly to eliminate the extra LiOH, and the ball-milled powders were collected by centrifugation, and finally dried under vacuum. The as-prepared BP-OH powder was dispersed in aqueous solution and centrifuged at 7000 rpm for 10 min. The supernatant of BP-OH was defined as several layers of BP-OH nanosheets. The as-prepared BP-OH powder was dispersed in aqueous solution and then centrifuged at 7000 rpm for 10 min. The supernatant of hydroxyl functionalized BP is defined as BP-OH nanosheets.

4.2.3 *Preparation of the BP-OHx/NFC Composite Films (X = 0, 5, 10, 25, 40)*

First, the NFC suspension was obtained by TEMPO-mediated oxidation method [17]. The dried exfoliation BP-OH was redispersed in N_2 saturated water with various concentration, follow by added the drops to the NFC suspension (0.5 wt %) for continuous stirring. Then, the suspension containing NFC and BP-OH was stirred and sonicated for 8 h under N_2 to generate a uniform blend mixture. The homogeneous mixture was poured into sand core funnel and filtered to form the BP-OHx/NFC composite film by the vacuum-assisted filtration. For comparison, pure NFC film was also fabricated used the same method under the same conditions. The weight ratio of BP-OH to NFC was 0:100, 5:95, 10:90, 25:75 and 40:60.

4.2.4 *Characterization*

Refer to Chaps. 2 and 3 for FTIR, XRD, Raman, XPS, TEM, SEM, AFM, TGA, DMA and MCC measurements.

4.3 Results and Discussion

4.3.1 *Morphology and Structure Characterization of BP-Bulk and BP-OH*

During the high-energy ball milling process, BP is unstable and easy to react. Selectively, we utilize LiOH powder as a modifier and mix with bulk BP powder, as a result, the reactive edges of BP nanosheets is terminated by –OH groups. Graphene is a two-dimensional layered structure similar to BP. During the mechanochemical cracking of graphene C–C bonds, the activated carbon (free radicals, ions) is formed at graphene edges, then the activated carbon on the edges of graphene nanosheets react with the reactants during ball-milling process, for example, the reactive groups including –COOH and –OH can be selected to form [24]. Referring to the edge-selective mechanism of modified graphene, thus, in a similar mechanochemical cleavage process, the reactive species like hydroxyl groups (–OH) can be formed in the BP-OH through cleavage of P-P bonds. In order to reveal the structure and composition of BP-OH relative to BP-Bulk, the morphological and spectral characterizations were carried out. In Fig. 4.2, from the IR spectrum of BP-OH, there are several typical peaks as same as the BP-Bulk, besides, it observed the peaks at 3465 cm^{-1} corresponds to the stretching vibration of -OH bonds [14].

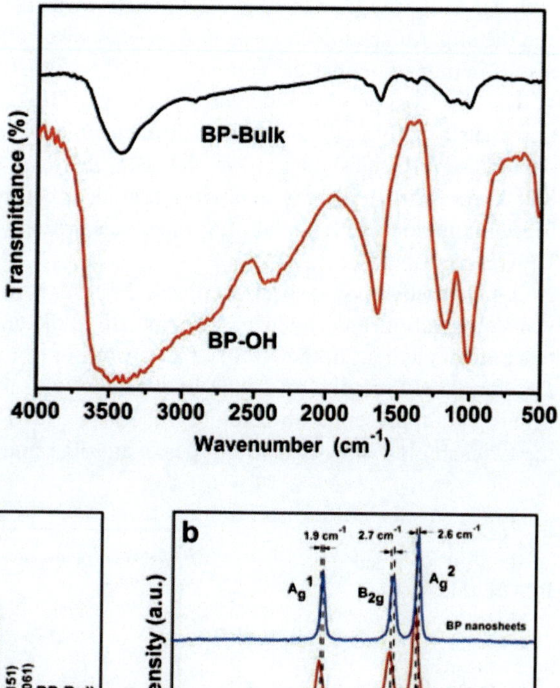

Fig. 4.2 FT-IR spectra of BP-bulk and BP-OH nanosheets

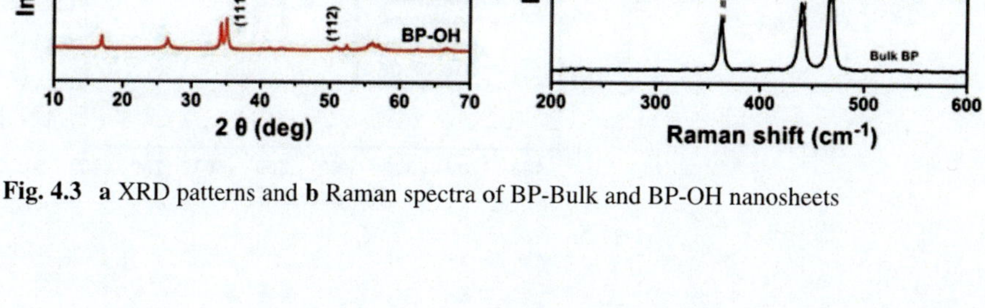

Fig. 4.3 **a** XRD patterns and **b** Raman spectra of BP-Bulk and BP-OH nanosheets

As shown in Fig. 4.3a, the XRD patterns of BP-Bulk and BP-OH well correspond to the orthorhombic phase of BP. The sharp diffraction peaks in the XRD pattern of BP-Bulk at $2\theta = 17.0°$, 34.4°, and 52.3° distinctly correspond to the (020), (040), and (060) planes, respectively, they are typical lamellar planes of the black phosphorus [25]. In addition, the (111) plane becomes the strongest diffraction peak rather than (040) plane of the BP-OH, it indicates that the crystal plane which parallel to the layered structure is disturbed, because of the interactions between nanolayers in BP-OH nanosheets become weaker and decrease the number of layers [26]. For BP-OH, its XRD peaks are broader and weaker, revealing that the crystallinity decreases and the grain size after ball-milling is relatively small.

As shown in Fig. 4.3b, the Raman spectrum of BP-Bulk shows several characteristic vibrational modes of BP-Bulk belong to the peaks of A_g^1 at 363.3 cm^{-1}, B_{2g} at 439.3 cm^{-1}, and A_g^2 at 467.8 cm^{-1}. It highlights that since the intensity ratio of A_g^1/A_g^2 is about 0.6, there is no oxidation of BP nano-sheet. In addition, three main peaks of BP-OH powder were also found, indicating that the vibration structure of BP with hydroxyl group remained after functionalization. Nonetheless, compared with BP-Bulk, the Raman spectra of the BP nanosheets show three stronger peaks, and the slight blue shift of these Raman peaks was observed (about 2 cm^{-1}). These results demonstrate that the nanolayer numbers of BP nanosheets are reduced [13].

The mass loss process and hydroxyl group of BP-OH were determined by thermogravimetric analysis (TGA) under nitrogen. The BP-Bulk has weight loss between 450 and 550 °C, as shown in Fig. 4.4. But, there are two steps of mass loss before 800 °C for BP-OH, there is an obvious mass loss before 250 °C, because of hydroxyl heat elimination and removal of physical adsorption water, and other defects on the BP-OH nanosheets.

X-ray photoemission spectroscopic (XPS) was carried out to investigate the chemical composition and valence states of BP-Bulk and BP-OH. There is no existence of Li element in the XPS survey spectrum of the BP-OH, demonstrating that the completely removal of LiOH by washing with N_2 saturated water (Fig. 4.5a). Figure 4.5b-d present high-resolution P 2p spectra of BP-Bulk and BP-OH, both of the two samples show two strong peaks at 130.1 and 130.9 eV. For BP-Bulk, its P

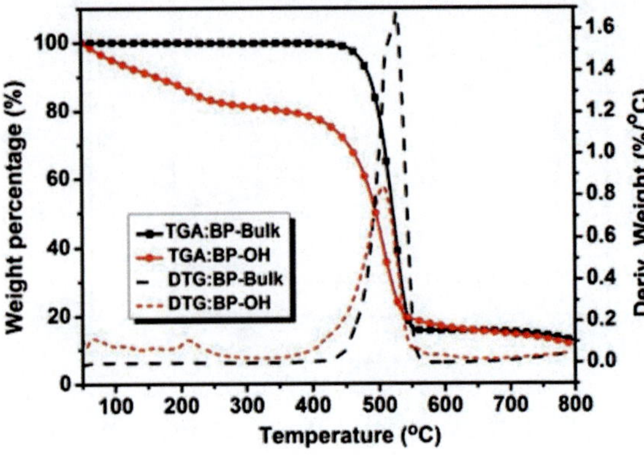

Fig. 4.4 TGA and DTG curves of BP-Bulk and BP-OH nanosheets

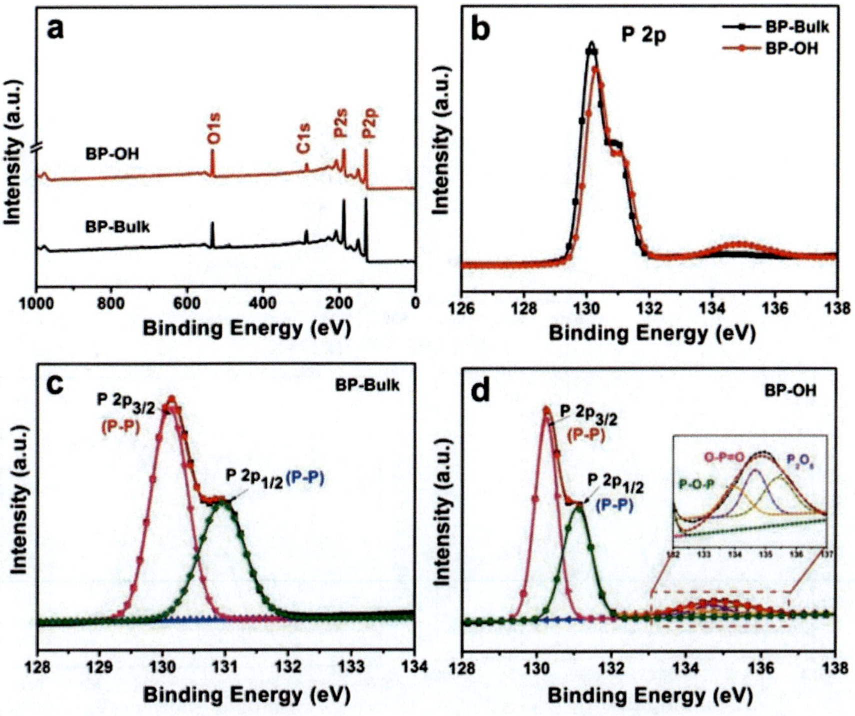

Fig. 4.5 XPS survey spectra of the BP-Bulk and BP-OH; **a–c** High-resolution P 2p XPS spectra of BP-bulk and BP-OH

2p peaks are deconvoluted into two peaks which are ascribed to P $2p_{3/2}$ and P $2p_{1/2}$ of P-P bonds. Besides, the P 2p spectra of BP-OH can be divided into five peaks at 130.1, 130.9, 133.9, 134.6 and 135.3 eV which are attributed to P-P bonds (P $2p_{3/2}$ and P $2p_{1/2}$), P–O–P and O–P = O bonding, and P_2O_5, respectively [27]. Among them, the relative strength ratio of P $2p_{1/2}$ to P $2p_{3/2}$ changes compared with that of BP-OH and BP-Bulk, revealing that BP-OH has structural defects due to the P–P bond cleavage at BP nanoflake edges during ball-milling procedure [28].

Figure 4.6a-c shows the high-resolution O1s XPS spectra which can also prove these results. For BP-Bulk, its O1s XPS spectrum can be deconvoluted into two strong peaks at 532.2 and 533.4 eV correspond to P=O (dangling bonding) and P–O–P (bridging bonding), respectively. Except for the P–O–P and P=O peaks, we can observed an extra P–OH bond at 534.0 eV for BP-OH [29], compared to the results of hydroxyl functionalized graphene formed by ball-milling process [30], revealing the bulk BP can be exfoliated into ultrathin BP nanoflakes during high-energy ball-milling process, meanwhile, BP nanosheets can be functionalized with -OH groups by addition of LiOH. The -OH group is more likely to be formed at the edge of the BP nanosheet rather than on the surface of BP. These results are certified by the lower relative intensity of the bridging bonding (P–O–P) signal (533.4 eV) in the O 1 s spectrum of BP-OH, than that of BP-Bulk.

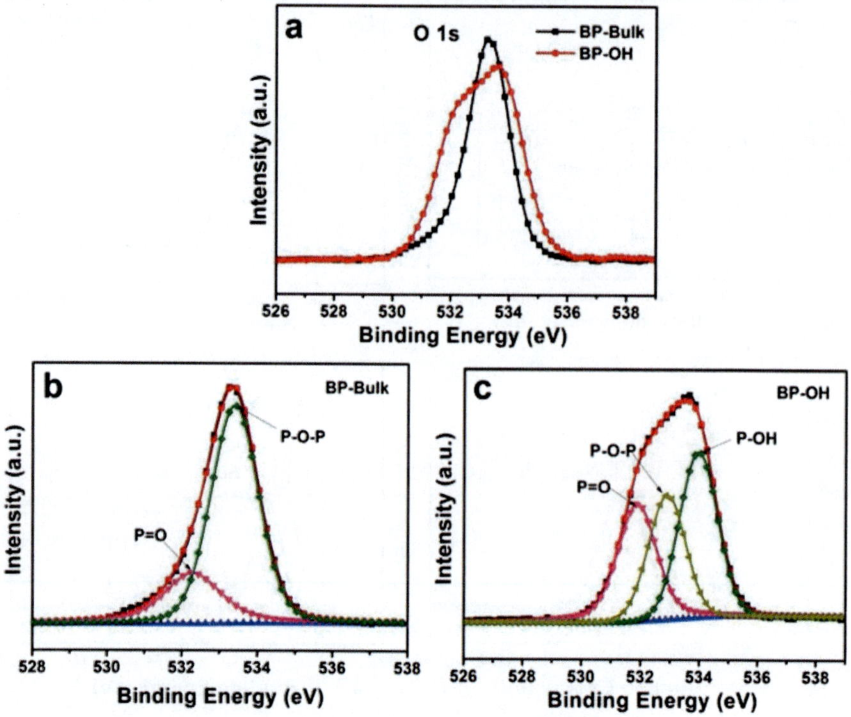

Fig. 4.6 **a–c** High-resolution O 1 s XPS spectra of BP-Bulk and BP-OH

The BP-OH was characterized by TEM and AFM. The bulk BP crystal has multilayer structure. As shown in the TEM images (Fig. 4.7a–d), the BP nanosheets stripped from BP-Bulk are large sheet morphology with a length of several microns and good crystallinity (Fig. 4.7a). The resultant BP-OH (Fig. 4.7b) maintains the well nanoflake morphology with minor defects after functionalized with hydroxyl group, and the size of BP-OH nanoflake is decrease compared to bulk BP nanoflake. The high-resolution TEM image (HRTEM) and the corresponding selected area electron diffraction pattern (SAED) in the inset of Fig. 4.7c illustrate the orthorhombic crystalline structure of the BP nanosheets. The lattice fringes of 0.25 and 0.34 nm are assigned to the (111) and (021) planes of BP nanosheets, respectively, consistent with XRD analysis results [31]. Figure 4.7d presents the atomic structure model of single-layer and folded multilayer BP nanoflakes.

Figure. 4.8a, b show the enlarged AFM image and height distribution of several BP-OH nanoflakes. The surface of BP-OH nanosheet was smooth with an average thickness of 3.27 and 5.56 nm, respectively. It is generally believed that the thickness of single layer of black phosphorus is 0.53 nm, thus, the thickness of BP nanosheets in Fig. 4.8a is 3.27–5.56 nm, which is equal to 5–9 layers of P atoms. The resultant lamellar BP can be defined as black phosphorene.

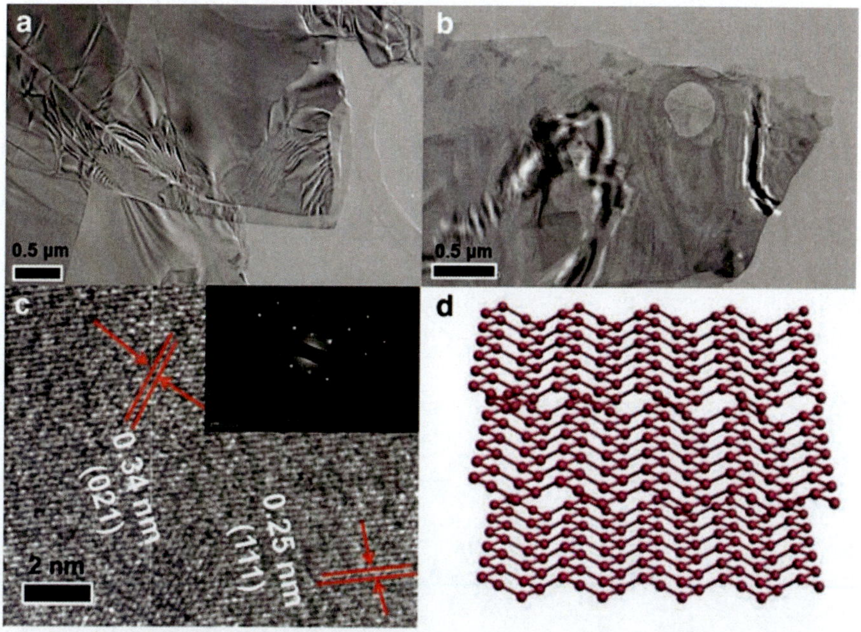

Fig. 4.7 TEM images of the **a** BP nanosheets and **b** BP-OH nanosheets; **c** HRTEM image and the corresponding SAED pattern of BP-OH nanosheets; **d** the atomic structure model of monolayer and folded-bilayer BP nanosheets

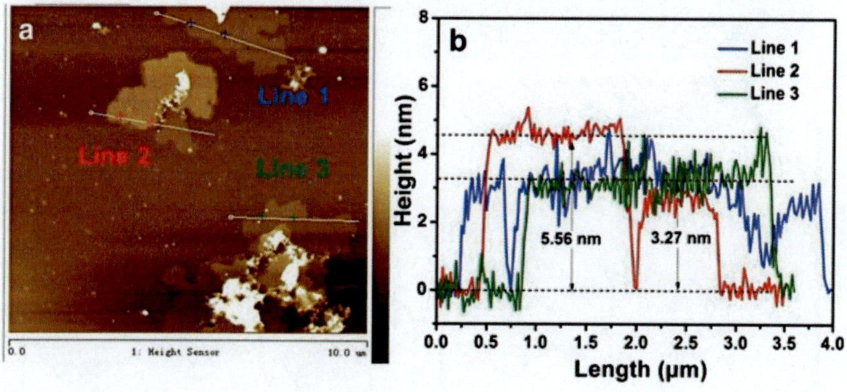

Fig. 4.8 **a** AFM image of BP-OH nanosheets and **b** the corresponding height profiles taken along the lines marked in (**a**)

4.3.2 *Morphology and Structure Characterization of BP-OHx/NFC*

The binary artificial nacre BP-OHx/NFC was built by assembling 1D and 2D building blocks with hydroxyl bonding. After the BP-OH suspension and NFC suspension were mixed for 8 h under stirring and sonication, numerous of small aggregates formed, as shown in Fig. 4.9a, revealing that there exhibit interfacial interactions between 1D NFC and 2D BP-OH nanosheets. After exfoliation and modification by ball milling process, the reactive nanosheet edge of BP is terminated by hydroxyl groups. As a consequence, NFC can readily attach on the BP-OH nanosheet surface,

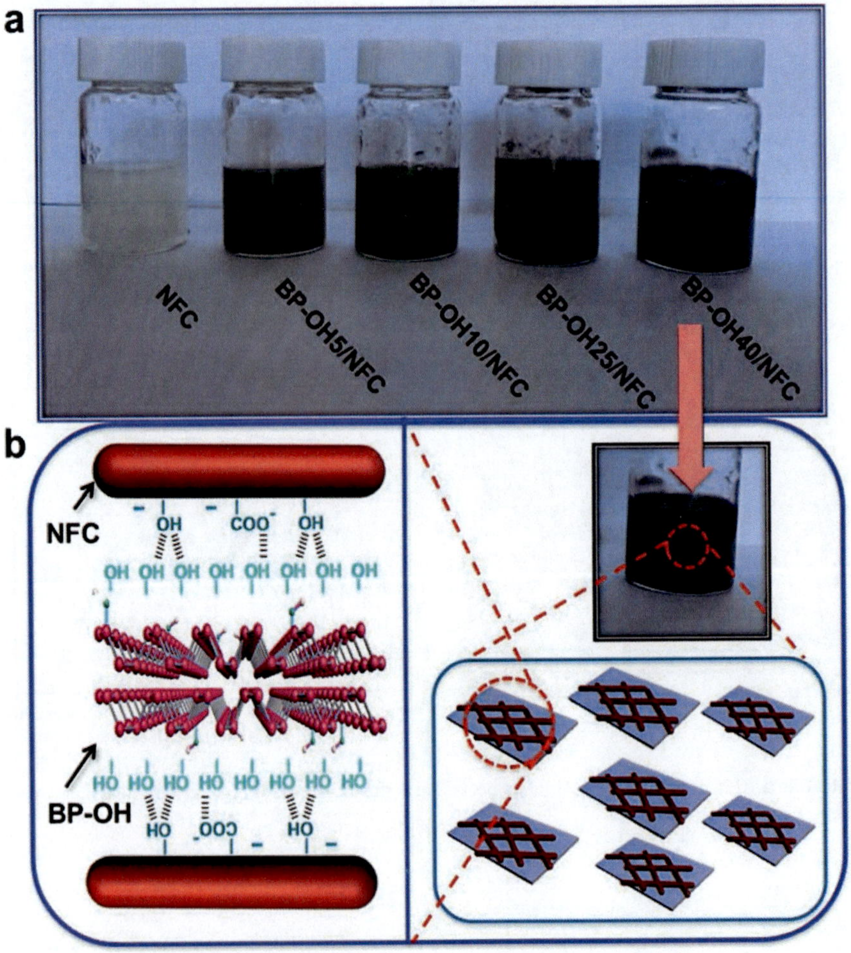

Fig. 4.9 **a** Digital image of BP-OHx/NFC suspension; **b** schematic illustration of the interaction between BP-OH and NFC by hydrogen bonds

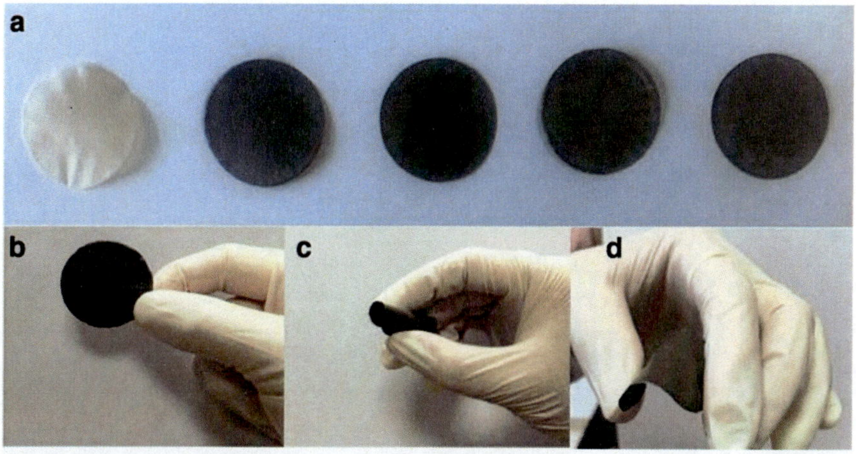

Fig. 4.10 **a–d** Digital images of the as-prepared pure NFC film and BP-OHx/NFC composite films

it can be believed that the construction of interface interactions between BP-OH nanosheets and NFC via hydrogen bonding network (Fig. 4.9b).

The digital image of the resultant binary bioinspired BP-OHx/NFC nanocomposite films are shown in Fig. 4.10a (Form left to right: 1 NFC, 2 BP-OH5/NFC, 3 BP-OH10/NFC, 4 BP-OH25/NFC, 5 BP-OH40/NFC), which show smooth surface with a bit of metallic luster. Figure 4.10b–d show the digital images of the BP-OH25/NFC sample, when the BP-OH25/NFC composite film is folded into two, there is no crack or fracture observed, the BP-OH25/NFC shows satisfied strength due to the synergistic toughening effect of BP-OH nanoflakes and NFC.

SEM was utilized to characterize the microstructures of the BP-OHx/NFC composite films, and we found that these composite films exhibit the "brick-and-mortar" structure, like the layered structure of nature nacre. As we know, the superior mechanical properties of natural nacre materials are derived from the hard and soft components of their hierarchical structure. We can develop the strategy of nanocomposites with high toughness and strength by imitating biological assembly process in nacre. In this work, a vacuum filtration strategy is adopted to imitate the bionic self-assembly procedure and presents a great potential in fabricating the BP-OHx/NFC nanocomposites with a nacre-like structure. The SEM images of the BP-OHx/NFC composite films show a "brick-and-mortar" structure, when the weight ratio of BP-OH in the BP-OHx/NFC varies from 10 to 40 w% (Fig. 4.11a–c). For the BP-OH10/NFC composite film, the BP-OH content is 10 wt%, which shows a tightly stacked layered structure. It can be seen that the directional alignment of BP-OH nanosheets along the planar direction are encompassed by cross-linking NFC. When the BP-OH content is 25 wt%, the BP-OH25/NFC composite film also presents a tightly stacked layered structure. With BP-OHx content increasing to 40 wt%, excess BP-OH nanosheets forming a wavy lamellar structure. In these nacre-excited structures, the two-dimensional BP-OH nanosheet plays the role of inorganic "brick",

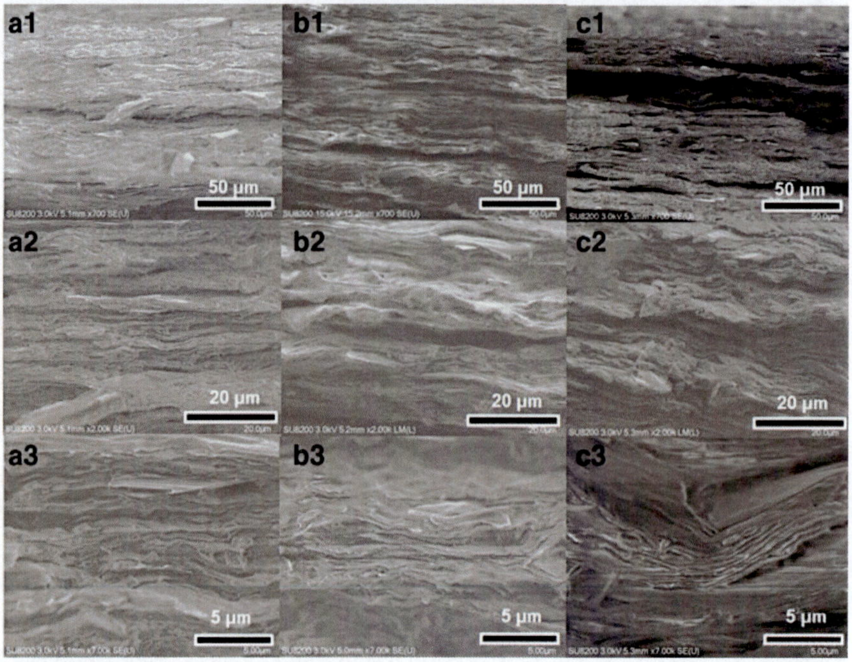

Fig. 4.11 SEM images of the BP-OH10/NFC (a1-3), BP-OH25/NFC (b1-3) and BP-OH40/NFC (c1-3) composite films in different magnification, exhibiting a nacre-like compact lamellar structure

and the one-dimensional cross-linked NFC plays the role of organic "mortar". This nacre-like biomimetic structure not only possesses superior mechanical properties but also is beneficial to apply as fire and heat insulation material.

Figure 4.12 presents SEM image of BP-OH25/NFC composite film (a–b) and relevant elemental mapping images (c-e). The mapping of carbon (C), oxygen (O) and phosphorus (P) further confirm the successful formation of BP-OH25/NFC composite.

The FTIR spectra of the pure NFC, BP-OH, and BP-OHx/NFC (x = 5, 10, 25, 40) composite films are displayed in Fig. 4.13. The BP-OH has two distinct peaks at 848 and 1180 cm^{-1}, correspond to the stretching vibration of P-O and P = O bonds, revealing the hydroxyl functionalization of bulk BP is successful [32]. For the pure NFC film, the cellulose characteristic absorption bands at 2920 cm^{-1} is ascribed to C-H stretching, besides, the absorption peaks at 1621 and 663 cm^{-1} corresponding to -OH bending and -OH out-of-plane bending, these typical absorption bands can be seen in the FTIR spectra of the BP-OHx/NFC composite films as well. Besides, compared to the NFC film, the typical bands of NFC part at 1621 and 3465 cm^{-1} are obviously shifted to high wave-number in the BP-OHx/NFC composite films, which indicating strong intermolecular hydrogen bonds form between the NFC and BP-OH due to those two nanomaterials both exhibit hydroxyl groups, not just simple physical recombination.

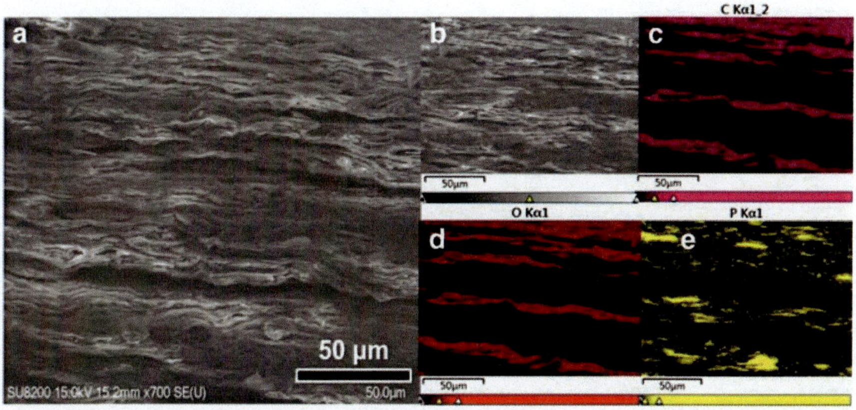

Fig. 4.12 **a** and **b** SEM image of BP-OH25/NFC composite film and corresponding elemental mapping images of **c** carbon (C), **d** oxygen (O) and **e** phosphorus (P)

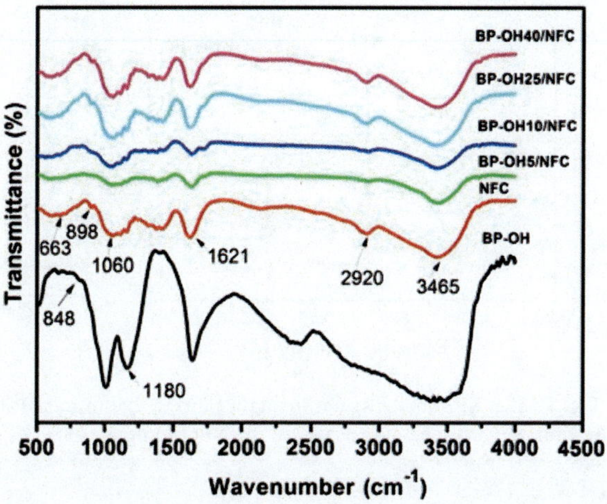

Fig. 4.13 FTIR spectra of the pure NFC film and BP-OHx/NFC composite films with different BP-OH contents

The XRD patterns of the BP-OHx/NFC composites are presented in Fig. 4.14. There exists overlapping peaks at $2\theta = 16.9°$ in the BP-OHx/NFC composite films are assigned to (101) planes of cellulose, Besides, the intense peak at $2\theta = 23°$ is ascribed to the (002) plane of crystalline structure of cellulose [33]. These two typical peaks of NFC become weaker with increasing loading of BP-OH from 5 to 40 wt% in BP-OHx/NFC composite films. Besides, the representative (002) peak slightly shift from $2\theta = 23$ to $22.7°$, indicating that the d-spacing of BP-OH nanosheets increases for the BP-OHx/NFC composite film, this phenomenon demonstrate that the intercalation of NFC into the nanolayers of BP-OH nanosheets.

Figure 4.15a presents XPS survey spectra of NFC and BP-OH25/NFC. For BP-OH25/NFC sample, it contains C, O and P elements, compared to pure NFC sample

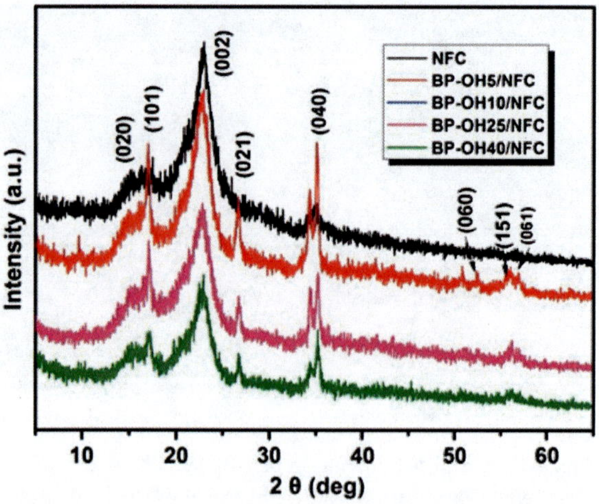

Fig. 4.14 XRD patterns of the pure NFC film and BP-OHx/NFC composite films with different BP-OH contents

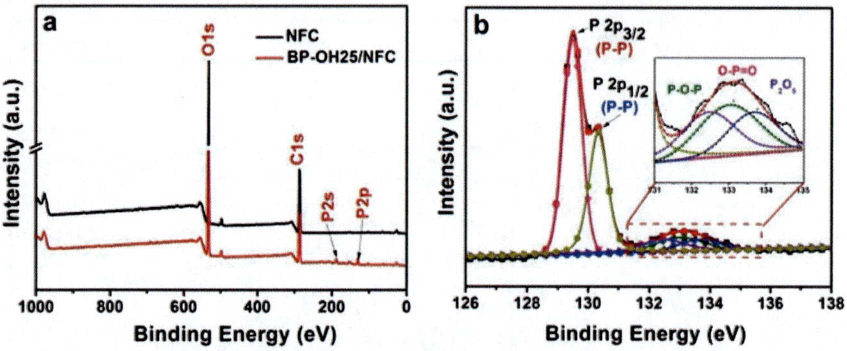

Fig. 4.15 **a** XPS survey spectrum of the pure NFC film and BP-OH25/NFC composite film; **b** High-resolution P 2p spectrum of BP-OH25/NFC composite film

only contains C and O elements. The high-resolution P 2p XPS spectrum of BP-OH25/NFC is shown in Fig. 4.15b. Fitting analysis demonstrates that the P 2p peaks of BP-OH25/NFC are decomposed into five peaks at 130.1, 130.9, 133.9, 134.6 and 135.3 eV, respectively, correspond to P-P bonds (P $2p_{3/2}$ and P $2p_{1/2}$), P-O-P and O-P=O bonds, and P_2O_5, which imply the existence of BP-OH in the BP-OH25/NFC composite film. All these results reveal that the BP-OH and NFC have been successfully integrated in the resulting BP-OHx/NFC composite films.

4.3.3 *Mechanical Properties of BP-OHx/NFC*

The tensile stress–strain curves of the BP-OHx/NFC (x = 5, 10, 25, 40) composite films are presented in Fig. 4.16a. The tensile tests indicate that the BP-OHx/NFC composite films show superior mechanical strength integration (Fig. 4.16b, c). The NFC film possesses a tensile strength and fracture strain of 52.0 MPa and 4.6%, respectively. The tensile strength of the BP-OHx/NFC composite films increases from 111.8 to 214.0 MPa and then decreases to 149.8 MPa, with increasing BP-OH loading from 5 to 40 wt%; meanwhile, the tensile fracture strain increases from 13.8 to 23.8%. When the BP-OH loading is 25 wt%, the optimal mechanical performance of the BP-OHx/NFC composite film is reached, and the relevant tensile strength and tensile fracture strain are 214.0 MPa and 18.7%, respectively. For BP-OHx/NFC binary nanocomposite films, the maximum tensile strength achieve at the BP-OH loading of 25 wt% and the maximum tensile fracture strain achieve at the BP-OH loading of 40 wt%. When the BP-OH content is increasing from 5 to 25 wt%, the cross-linking action includes covalent bonding and hydrogen bonding between BP-OH and NFC is gradually increasing, can enhance the strength and interfacial interaction of the BP-OHx/NFC composite films; when the BP-OH content is over

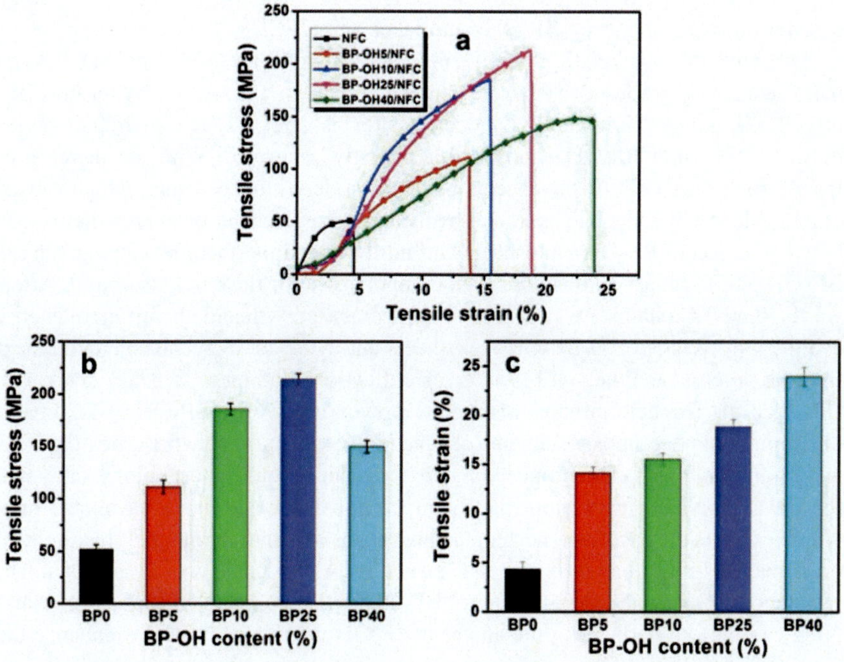

Fig. 4.16 Mechanical properties of the pure NFC film and BP-OHx/NFC composite films with different BP-OH contents. **a** Tensile stress–strain curves of the pure NFC film and BP-OHx/NFC composite films with different BP-OH contents; **b** tensile strengths and **c** tensile strains data of the pure NFC film and BP-OHx/NFC composite films with different BP-OH contents

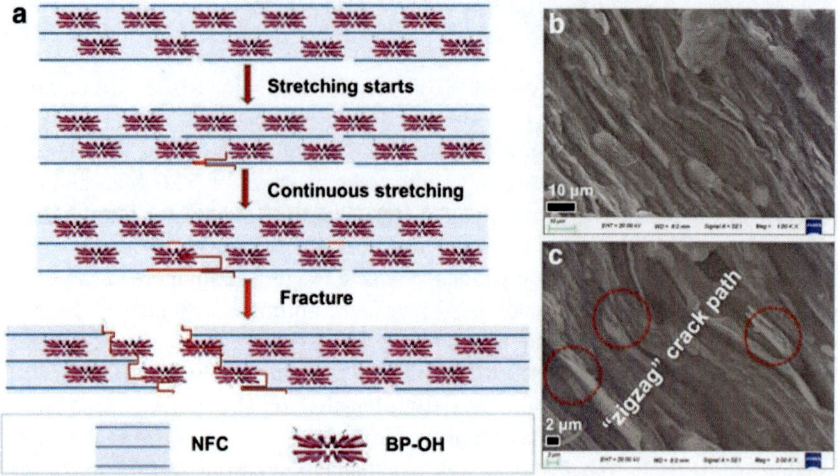

Fig. 4.17 **a** Schematic illustration of the fracture mechanism of the BP-OHx/NFC composite films; **b** and **c** SEM micrographs of the fracture surface of the BP-OHx/NFC composite films

25 wt%, BP-OH was completely surrounded by NFC, leading to reducing the energy transfer efficiency between BP-OH nanoflakes and NFC.

To know the synergistic strengthening effect from 1D NFC and 2D BP-OH nanosheets, we proposed a crack extension model in Fig. 4.17a. When the BP-OHx/NFC binary bioinspired nanocomposites is stressed, the hydrogen bonds between NFC and BP-OH nanosheets are firstly destroyed, With the increase of tensile force, the BP-OH nanosheets began to slide with each other, which caused cracks. Meanwhile, the NFC part can withstand excessive friction energy dissipation between adjacent BP-OH nanosheets and inhibit the slip effect, thus strengthen the BP-OHx/NFC binary bioinspired nanocomposites until the cracks forming. Afterwards, the NFC chains are stretched to slow crack growth and absorb more energy as they are stretched. 2D BP-OH nanosheets and NFC act a crucial role in the next fracture process, and they will form crack deflection, leading to "zigzag" crack path (Fig. 4.17c). The determinants of fracture prevention for the BP-OHx/NFC binary bioinspired nanocomposites are anisotropic interconnection networks, flexible NFC and hydrogen bonds. The proposed mechanism and fracture morphology show that the BP-OHx/NFC binary bioinspired nanocomposite seems like a nacre structure, which exhibits the excellent mechanical properties originated from their hierarchical organization of the "brick-and-mortar" structure. As "bricks", two-dimensional BP-OH nanoflakes provide the skeleton for BP-OHx/NFC composite films; as "mortar", NFC can not only adhere to the adjacent BP-OH nanoflakes, but also enhance the friction energy dissipation and improve the stress transfer efficiency. Figure 4.17b–c display the SEM images of BP-OHx/NFC composite film after the fracture tensile test, we can observe that the fracture surface still maintains a tightly stacked lamellar structure, similar to the structure of the nacre, which is integrated by the interaction between1D NFC and 2D BP-OH nanosheets in the process of vacuum filter-induced self-assembly. The existence of BP-OH nanosheets and NFC polymer are observed in the SEM image. Besides, the rough fracture surface of BP-OH25/NFC film shows obvious "zigzag" crack paths and crack deflection.

The mechanical and thermal performances of composite films were enhanced after incorporating BP-OH nanosheets into NFC matrix, these results were comfirmed by DMA test. The storage modulus and tan δ of the BP-OHx/NFC composite films containing 5, 10, 25 and 40% as a function of temperature are shown in Fig. 4.18. The storage modulus of pure NFC films at room temperature is about 2238.68 MPa. In detail, the storage modulus of BP-OHx/NFC composite films at room temperature are dramatically improved by 139.1%, 162.9%, 80.7% and 52.1%, respectively, with the increase of BP-OH contents from 5 to 40 wt%, compared to that of pure NFC film. The significant increase of storage modulus is mainly ascribed to the hydrogen bonding and interfacial interaction between BP-OH and NFC, and the high stiffness of BP-OH. Particularly, the DMA data for BP-OH10/NFC composite film exhibits a storage modulus of 5885.37 MPa at room temperature. As expected, the storage modulus even reaches 5022.2 MPa until 220 °C. The BP-OHx/NFC films show higher modulus thermal stability which superior to typical polymer materials or nanocomposites. The explanation is that the inherent characteristics of BP-OH and mechanical strength of NFC, combined with ordered structure of their composites. In the tan δ curves, the peak values are assigned to glass transition temperature (T_g) of the BP-OHx/NFC composite films (Fig. 4.18b). Concretely, with the introducing of different loadings of BP-OHx (x = 5, 10, 25, 40) into composite films, the T_g values of composite films are slightly increased than that of pure NFC. The increase of T_g value demonstrates that the improvement of the interfacial interaction between the NFC chains and the surface of BP-OH nanoflakes, thereby enhancing the mechanical and thermal properties of BP-OHx/NFC composites.

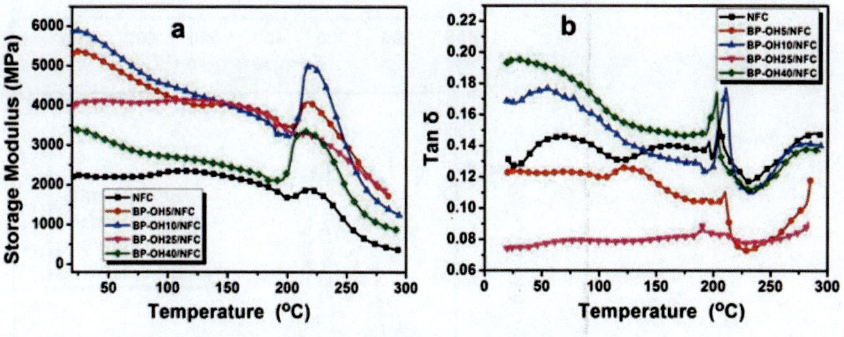

Fig. 4.18 **a** Storage modulus (E′) curves and **b** Tan δ curves of the pure NFC film and BP-OHx/NFC composite films with different BP-OH contents as a function of temperature

4.3.4 Thermal Stability of BP-OHx/NFC

Thermogravimetric analysis (TGA) was used to study the effect of BP-OH nanosheet on the thermal properties of BP-OHx/NFC composite films under nitrogen (N_2). Upon the Fig. 4.19a, BP-OHx/NFC composite films show a three-stage weight loss procedure, are separated into thermal cracking at 200–280 °C and 280–350 °C and carbonization at 400–500 °C, the pyrolysis behavior is similar to that of pure NFC. The decomposition temperature at the region of 200–300 °C of the BP-OHx/NFC composite films increased obviously. After introducing the BP-OH with different contents into NFC, the char residues of BP-OHx/NFC composites at 800 °C are gradually increased significantly, compared to that of pure NFC, due to the superior thermal stability and catalytic charring effect of BP-OH. Besides, the maximum decomposition rate for three stages decreased with increasing of BP-OH loading, and high BP-OH content is beneficial for this phenomenon. The DTG curves (Fig. 4.19b) show that the decomposition rates of BP-OHx/NFC composites films are significantly lower than that of NFC film. As an inorganic part, BP-OH nanosheets can inhibit the

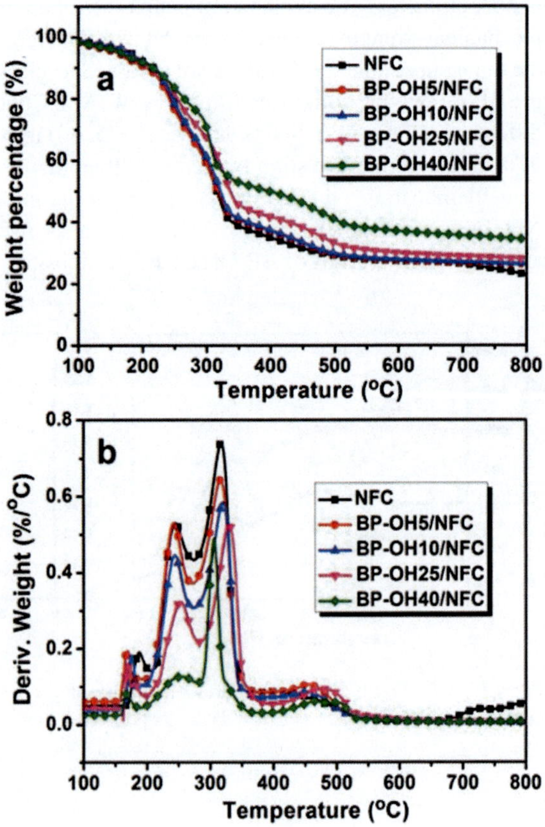

Fig. 4.19 **a** TGA curves and **d** DTG curves of the pure NFC film and BP-OHx/NFC composite films with different BP-OH contents

transport of oxygen and the release of pyrolysis products as a physical barrier, thus improving the anti-degradation thermal resistance of the composite films.

4.3.5 Fire Resistance of BP-OHx/NFC

The fire performances of BP-OHx/NFC composite films were studied by MCC test. The heat release rate (HRR) and total heat release rate (THR) vs. time curves of BP-OHx/NFC composite films are shown in Fig. 4.20a, b. Pure NFC burns violently with high PHRR and THR values of 93.0 W/g and 10.8 kJ/g, respectively. With increasing BP-OH content from 5 to 40 wt%, the PHRR values of the BP-OHx/NFC (x = 5, 10, 25, 40) composite films decreased by 17.7–46.9%; in addition, the THR values of the BP-OHx/NFC (x = 5, 10, 25, 40) composite films decreased by 16.6–34.6%. In particular, when 40 wt% of BP-OH was added into BP-OH40/NFC, the

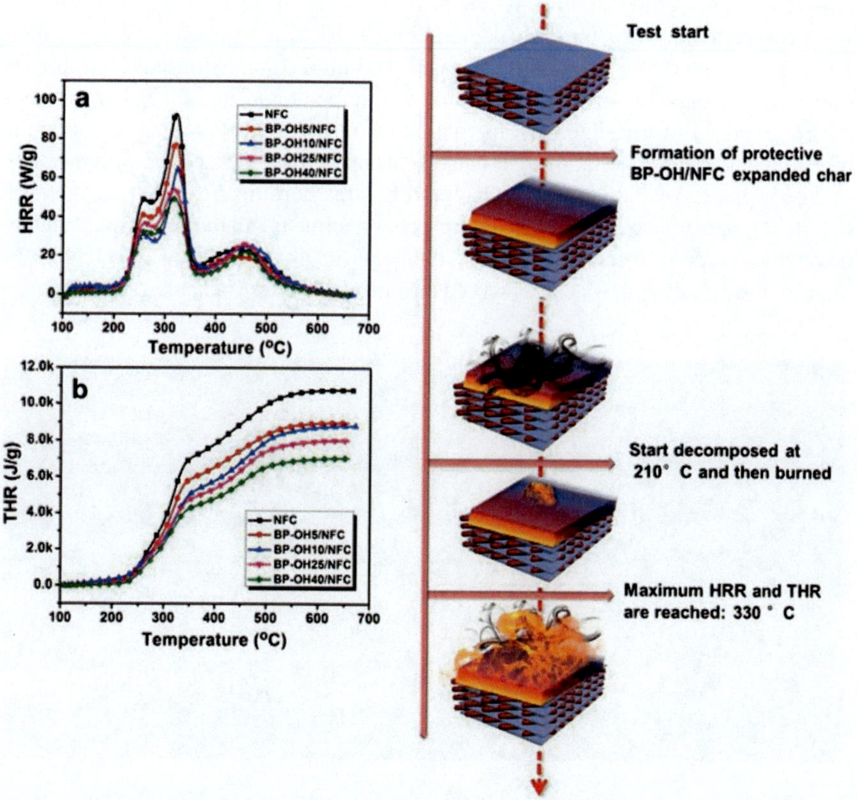

Fig. 4.20 **a** HRR and **b** THR vs. temperature curves of the pure NFC film and BP-OHx/NFC composite films with different BP-OH contents

maximum decrease of PHRR and THR during combustion was 46.9% and 34.6%, respectively, indicating that the flame retardant of BP-OHx/NFC composite films was improved with the increase of BP-OH loadings. These apparent improvements in the flame retardant performance of BP-OHx/NFC composite films can be ascribed to the physical barrier and catalytic charring effect of BP-OH nanoflakes [34].

Like the red phosphorus (RP), BP nanosheets may have good fire retardant properties due to the presence of layered structure and phosphorus elements [35]. Howerever, the BP nanosheets are seldom used for preparing flame retardant composites. The BP-OHx/NFC binary bioinspired nanocomposites show outstanding flame retardant performance with a certain amount of BP-OH which exist in the as-prepared nanocomposites. Besides, these binary bioinspired nanocomposites possess high thermal stability, which is proved by the TGA analysis as shown in Fig. 4.19. Hence, these BP-OHx/NFC composite films can be applied for the protection of flammable materials. As shown in Fig. 4.21a, b, a cotton ball behind BP-OH40/NFC composite film cannot ignite even upon prolonged exposure for more than 2 min, however, the cotton ball was burned rapidly within 5 s without the protection of the bioinspired BP-OH40/NFC composite film. When the bioinspired BP-OHx/NFC composite film was exposed to open fire, the early degradation of NFC and residual oxygen groups of BP make it produce a small amount of gas in the initial combustion stage. After a few minutes, the shape of the sample became a little bend and no longer burning even further exposed to fire, due to construction of an inorganic framework containing high content of orderly stacked BP nanosheets and crosslinking NFC polymer.

SEM images of NFC and BP-OH40/NFC after burning treatment as shown in Fig. 4.22a, b, for the pure NFC, its interior char residue is relatively compact and has no obvious layered structure. However, in the interior char residue of BP-OH40/NFC after burning treatment, we can observed the porous structure because of the diffusion

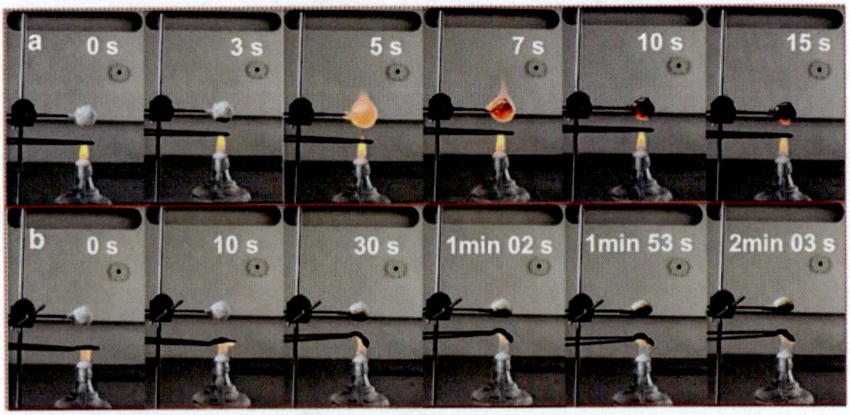

Fig. 4.21 Binary bioinspired nanocomposites act as a fire shield to protect a cotton ball. **a** It took only 5 s to burn the cotton ball. **b** The cotton ball is well kept from burning for more than 2 min with BP-OH40/NFC binary bioinspired nanocomposites

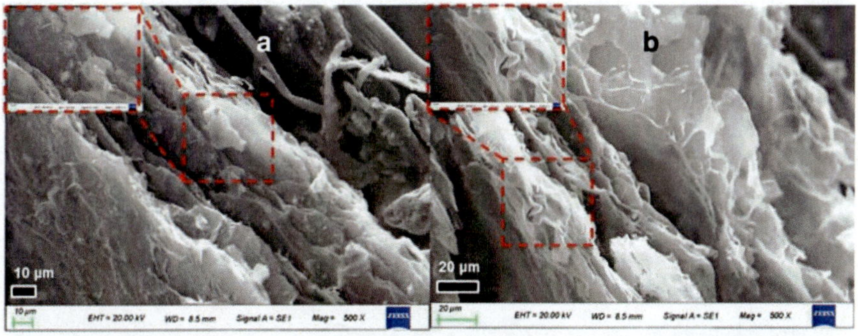

Fig. 4.22 SEM images of the NFC **a** and BP-OH40/NFC **b** binary bioinspired nanocomposites after flame treatment, showing the tightly armored skin and mesoporously layered interior

of volatile products from NFC and residual oxygen groups of BP-OH; simultaneously, the binary bioinspired nanocomposites remain their layered structure due to the high thermal and flame resistance properties.

4.3.6 Flame Retardant Mechanism of BP-OHx/NFC

On the basis of thermal degradation and flammability data in previous sections have provided, a probable fire resistance mechanism of BP-OHx/NFC nanocomposite film is proposed in Fig. 4.23. As consider the BP-OHx/NFC show "brick-and-mortar" layered structure with excellent thermal resistance, the BP-OHx/NFC are perpendicular to the heat or fire direction, where the BP nanoplatelets are parallel to the composite film surface, the physical barrier effects are formed from the oriented BP nanosheets constituent. On the other hand, the cellulose phase should be considered, the nanofibrillar cellulose with extended chain crystallite structure can form thermally stable ladder polymers during carbonization which be able to fabricate carbon fibers [36]. The existence of oriented BP has an effect on degradation behavior of cellulose in different ways. Physical barrier effect of oriented BP nanoflakes results in slow rise in local temperature, and forms a locally oxygen-depleted environment during exposure to fire. During the combustion, the BP has been predominantly oxidized to various PO_x and phosphoric acid derivatives when it start to decomposed nearly at ~450 °C in air, similar to red phosphorus [37]. Thus, the oriented BP nanosheets near to the cellulose could catalyze charring formation reaction and thus generated thermally stable char. BP nanosheets play as physical barrier to hinder the flammable volatiles escape from cellulose leading to microporous structure formation, thereby reduce the thermal conductivity of this BP-OHx/NFC composite film. In BP-OHx/NFC nanocomposite film with high BP-OH loading, abundant BP nanosheets on the surface of film will promote charring effect of cellulose during

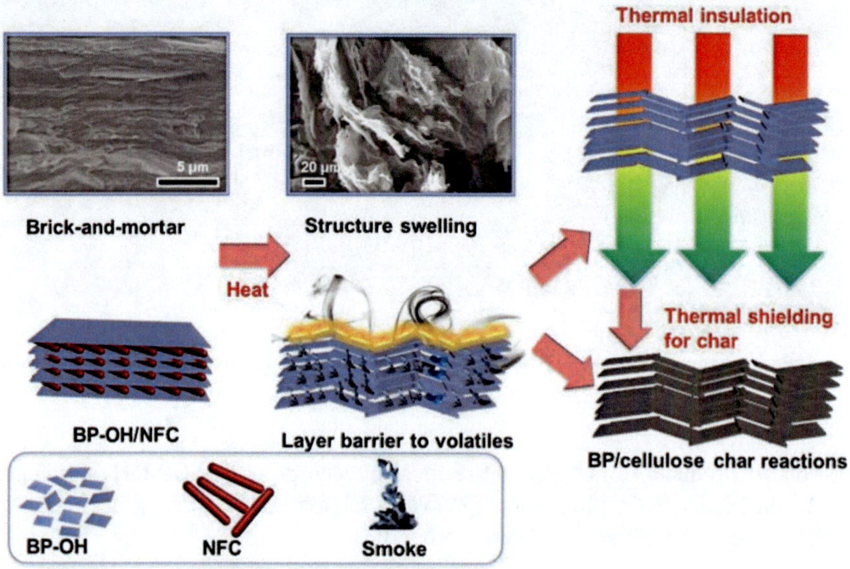

Fig. 4.23 Schematic sketch of the fire resistance mechanism attributed to the BP-OHx/NFC nanocomposite film

combustion. As a consequence, the BP-OHx/NFC nanocomposite film with "brick-and-mortar" layered structure has excellent fire resistance properties due to high BP content (as same as P element content), better nanosheet orientation, large interfacial area and volatile inhibiting effects.

4.3.7 Air Stability of BP–OHx/NFC

To evaluate the air stability of BP-OHx/NFC nanocomposite films, the BP-OH25/NFC nanocomposite films was exposed to atmospheric conditions for four months. The microstructure of fracture surface of BP-OH25/NFC sample was investigated by SEM. The composite papers have ordered lamellar structure as shown in Fig. 4.24a, which is similar to the "brick-and-mortar" structure of nacre, and the "pull-out" BP nanosheets in the BP-OH25/NFC nanocomposites can be clearly observed. Figure 4.24b–e show the SEM image of BP-OH25/NFC composite film and corresponding elemental mapping images of phosphorus (P), oxygen (O), and carbon (C). These results demonstrate that the BP-OH nanosheets remain in BP-OH/NFC nanocomposite films exhibit good air stability.

To evaluate the air stability of BP-OH/NFC nanocomposite films, the BP-OH25/NFC nanocomposite film was exposed to atmospheric conditions for four months. These BP-OH25/NFC sample was further investigated by XRD pattern (Fig. 4.25a) and Raman spectrum (Fig. 4.25b). Its XRD pattern exhibits three intense

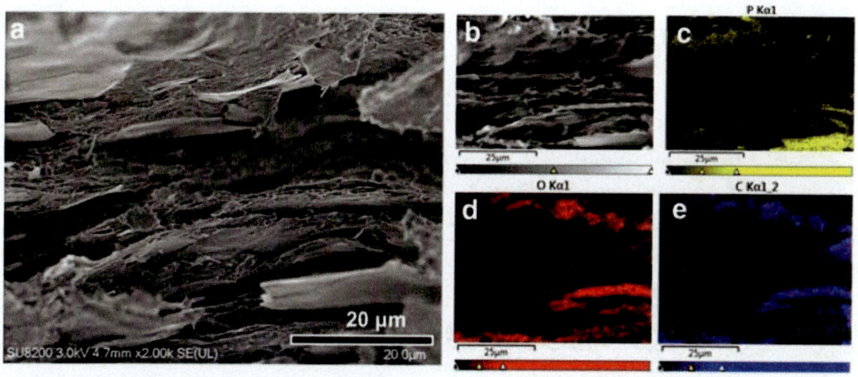

Fig. 4.24 **a** and **b** The SEM image obtained from BP-OH25/NFC nanocomposite film and corresponding elemental mapping images of **c** phosphorus (P), **d** oxygen (O) and **e** carbon (C) after exposure to ambient conditions for four months

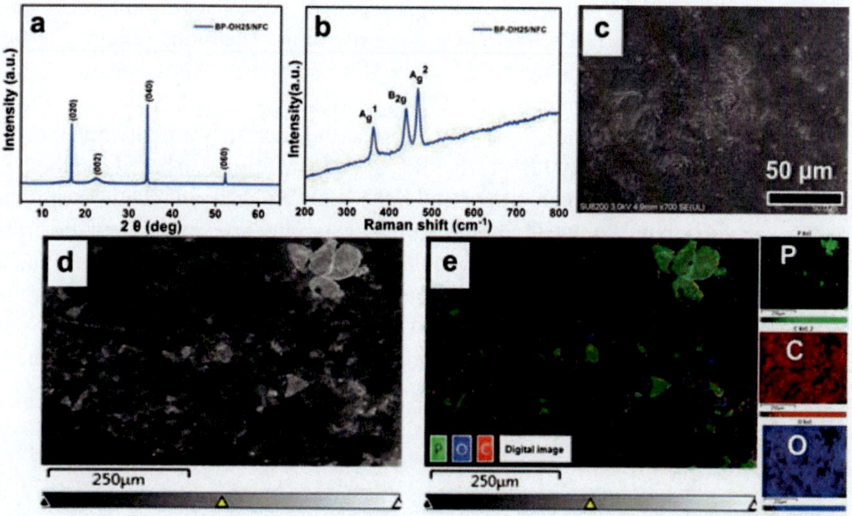

Fig. 4.25 The corresponding **a** XRD pattern and **b** Raman spectrum of BP-OH25/NFC nanocomposite film after exposure to ambient conditions for four months; **c** and **d** The SEM images obtained from the surface of BP-OH25/NFC nanocomposite film and corresponding **e** elemental mapping images of phosphorus (P), oxygen (O) and carbon (C) after exposure to ambient conditions for four months

diffraction peaks at 16.9°, 34.2° and 52.3 corresponding to the (020), (040), and (060) planes, respectively, which are representative layered planes of BP. the diffraction peak at $2\theta = 22.7°$ corresponds to the (002) crystal plane of cellulose crystalline structure. In the Raman spectrum, three characteristic vibrational modes of BP crystal correspond to the peaks of $A_g^{\ 1}$ at 360.7 cm^{-1}, B_{2g} at 438.6 cm^{-1}, and

A_g^2 at 466.4 cm^{-1}. These results confirm that the BP-OH in BP-OH/NFC nanocomposites with complete structure and crystallinity and cannot be degraded even on prolonged exposure to the air more than four months. The uniformly distributed crystalline BP nanosheets in the BP-OH/NFC nanocomposites can be clearly seen by SEM images and elemental mapping images as shown in Fig. 4.25c–e. As a consequence, these characterizations demonstrate that the BP-OH nanosheets in BP-OH/NFC nanocomposite films exhibit good air stability.

4.4 Summary

In this work, inspired by the "brick-and-mortar" structure of natural nacre, we utilized vacuum-assisted filtration self-assembly procedure to fabricate series of ultrathin and flexible BP-OHx/NFC composite films with a nacre-like microstructure. The optimized fracture tensile strength and tensile strain of the BP-OHx/NFC composite films can achieve 214.0 MPa and 23.8%, respectively, by changing the mass ratio of BP-OH to NFC in the BP-OHx/NFC composite films. The bioinspired nacre-like hierarchical structure forms in the BP-OHx/NFC composite films because of the strong interfacial interaction between 1D NFC and 2D few-layer BP-OH nanosheets via hydrogen bonding network, revealing a "brick-and-mortar" strengthening mechanism. In addition, the resultant BP-OHx/NFC composite films also have high thermal and fire retardant performance. Importantly, the XRD and Raman results confirmed that the BP-OHx/NFC composite films exhibit air stability after exposure to environmental conditions for 4 months. These multifunctional binary bioinspired composite films show great potential applications in the fields of flexible construction materials, as aerospace and flame retardant insulation materials. Such technique proposed here offers a novel innovative strategy for the preparation of multi-functional biomimetic nanocomposites by synergetic enhancement.

References

1. Zhang Y, Gong S, Zhang Q, Ming P, Wan S, Peng J, Jiang L, Cheng Q (2016) Graphene-based artificial nacre nanocomposites. Chem Soc Rev 45:2378–2395
2. Cheng Q, Duan J, Zhang Q, Jiang L (2015) Learning from nature: constructing integrated graphene-based artificial nacre. ACS Nano 9:2231–2234
3. Shahzadi K, Zhang X, Mohsin I, Ge X, Jiang Y, Peng H, Liu H, Li H, Mu X (2017) Reduced graphene oxide/alumina, a good accelerant for cellulose-based artificial nacre with excellent mechanical, barrier, and conductive properties. ACS Nano 11:5717–5725
4. Song P, Xu Z, Dargusch MS, Chen ZG, Wang H, Guo Q (2017) Granular nanostructure: a facile biomimetic strategy for the design of supertough polymeric materials with high ductility and strength. Adv Mater 29:1704661
5. Cheng Q, Jiang L, Tang Z (2014) Bioinspired layered materials with superior mechanical performance. Acc Chem Res 47:1256–1266

6. Ryder CR, Wood JD, Wells SA, Yang Y, Jariwala D, Marks TJ, Schatz GC, Hersam MC (2016) Covalent functionalization and passivation of exfoliated black phosphorus via aryl diazonium chemistry. Nat Chem 8:597
7. Tao W, Zhu X, Yu X, Zeng X, Xiao Q, Zhang X, Ji X, Wang X, Shi J, Zhang H (2017) Black phosphorus nanosheets as a robust delivery platform for cancer theranostics. Adv Mater 29:1603276
8. Ahmed F, Kim YD, Choi MS, Liu X, Qu D, Yang Z, Hu J, Herman IP, Hone J, Yoo WJ (2017) High electric field carrier transport and power dissipation in multilayer black phosphorus field effect transistor with dielectric engineering. Adv Func Mater 27:1604025
9. Cao X, Tan C, Zhang X, Zhao W, Zhang H (2016) Solution-processed two-dimensional metal dichalcogenide-based nanomaterials for energy storage and conversion. Adv Mater 28:6167–6196
10. Kwon H, Seo SW, Kim TG, Lee ES, Lanh PT, Yang S, Ryu S, Kim JW (2016) Ultrathin and flat layer black phosphorus fabricated by reactive oxygen and water rinse. ACS Nano 10:8723–8731
11. Kang J, Wood JD, Wells SA, Lee J-H, Liu X, Chen K-S, Hersam MC (2015) Solvent exfoliation of electronic-grade, two-dimensional black phosphorus. ACS Nano 9:3596–3604
12. Ni H, Liu X, Cheng Q (2018) A new strategy for air-stable black phosphorus reinforced PVA nanocomposites. J Mater Chem A 6:7142–7147
13. Shao L, Sun H, Miao L, Chen X, Han M, Sun J, Liu S, Li L, Cheng F, Chen J (2018) Facile preparation of NH 2-functionalized black phosphorene for the electrocatalytic hydrogen evolution reaction. J Mater Chem A 6:2494–2499
14. Zhu X, Zhang T, Sun Z, Chen H, Guan J, Chen X, Ji H, Du P, Yang S (2017) Black phosphorus revisited: a missing metal-free elemental photocatalyst for visible light hydrogen evolution. Adv Mater 29:1605776
15. Yao C, Yin X, Yu Y, Cai Z, Wang X (2017) Chemically functionalized natural cellulose materials for effective triboelectric nanogenerator development. Adv Func Mater 27:1700794
16. Zong Y, Zheng T, Martins P, Lanceros-Mendez S, Yue Z, Higgins MJ (2017) Cellulose-based magnetoelectric composites. Nat Commun 8:38
17. Peyre J, Pääkkönen T, Reza M, Kontturi E (2015) Simultaneous preparation of cellulose nanocrystals and micron-sized porous colloidal particles of cellulose by TEMPO-mediated oxidation. Green Chem 17:808–811
18. Toivonen MS, Kaskela A, Rojas OJ, Kauppinen EI, Ikkala O (2015) Ambient-dried cellulose nanofibril aerogel membranes with high tensile strength and their use for aerosol collection and templates for transparent, flexible devices. Adv Func Mater 25:6618–6626
19. Zeng X, Sun J, Yao Y, Sun R, Xu J-B, Wong C-P (2017) A combination of boron nitride nanotubes and cellulose nanofibers for the preparation of a nanocomposite with high thermal conductivity. ACS Nano 11:5167–5178
20. Jung YH, Chang T-H, Zhang H, Yao C, Zheng Q, Yang VW, Mi H, Kim M, Cho SJ, Park D-W (2015) High-performance green flexible electronics based on biodegradable cellulose nanofibril paper. Nat Commun 6:7170
21. Cao W-T, Chen F-F, Zhu Y-J, Zhang Y-G, Jiang Y-Y, Ma M-G, Chen F (2018) Binary strengthening and toughening of mxene/cellulose nanofiber composite paper with nacre-inspired structure and superior electromagnetic interference shielding properties. ACS Nano 12:4583–4593
22. Duan J, Gong S, Gao Y, Xie X, Jiang L, Cheng Q (2016) Bioinspired ternary artificial nacre nanocomposites based on reduced graphene oxide and nanofibrillar cellulose. ACS Appl Mater Interf 8:10545–10550
23. Jackson A, Vincent JF, Turner R (1988) The mechanical design of nacre. Proc R Soc Lond B 234:415–440
24. Jeon IY, Bae SY, Seo JM, Baek JB (2015) Scalable production of edge-functionalized graphene nanoplatelets via mechanochemical ball-milling. Adv Func Mater 25:6961–6975
25. Appalakondaiah S, Vaitheeswaran G, Lebegue S, Christensen NE, Svane A (2012) Effect of van der Waals interactions on the structural and elastic properties of black phosphorus. Phys Rev B 86:035105

26. Guan J, Chen X, Wei T, Liu F, Wang S, Yang Q, Lu Y, Yang S (2015) Directly bonded hybrid of graphene nanoplatelets and fullerene: facile solid-state mechanochemical synthesis and application as carbon-based electrocatalyst for oxygen reduction reaction. J Mater Chem A 3:4139–4146
27. Edmonds M, Tadich A, Carvalho A, Ziletti A, O'Donnell K, Koenig S, Coker D, Ozyilmaz B, Neto AC, Fuhrer M (2015) Creating a stable oxide at the surface of black phosphorus. ACS Appl Mater Interf 7:14557–14562
28. Liang L, Wang J, Lin W, Sumpter BG, Meunier V, Pan M (2014) Electronic bandgap and edge reconstruction in phosphorene materials. Nano Lett 14:6400–6406
29. Puziy A, Poddubnaya O, Socha R, Gurgul J, Wisniewski M (2008) XPS and NMR studies of phosphoric acid activated carbons. Carbon 46:2113–2123
30. Jeon I-Y, Choi H-J, Jung S-M, Seo J-M, Kim M-J, Dai L, Baek J-B (2012) Large-scale production of edge-selectively functionalized graphene nanoplatelets via ball milling and their use as metal-free electrocatalysts for oxygen reduction reaction. J Am Chem Soc 135:1386–1393
31. Zhang X, Xie H, Liu Z, Tan C, Luo Z, Li H, Lin J, Sun L, Chen W, Xu Z (2015) Black phosphorus quantum dots. Angew Chem Int Ed 54:3653–3657
32. Wood JD, Wells SA, Jariwala D, Chen K-S, Cho E, Sangwan VK, Liu X, Lauhon LJ, Marks TJ, Hersam MC (2014) Effective passivation of exfoliated black phosphorus transistors against ambient degradation. Nano Lett 14:6964–6970
33. Zheng Q, Cai Z, Ma Z, Gong S (2015) Cellulose nanofibril/reduced graphene oxide/carbon nanotube hybrid aerogels for highly flexible and all-solid-state supercapacitors. ACS Appl Mater Interf 7:3263–3271
34. Qiu S, Xing W, Mu X, Feng X, Ma C, Yuen RK, Hu Y (2016) A 3D nanostructure based on transition-metal phosphide decorated heteroatom-doped mesoporous nanospheres interconnected with graphene: synthesis and applications. ACS Appl Mater Interf 8:32528–32540
35. Laoutid F, Ferry L, Lopez-Cuesta J, Crespy A (2003) Red phosphorus/aluminium oxide compositions as flame retardants in recycled poly (ethylene terephthalate). Polym Degrad Stabil 82:357–363
36. Wu ZY, Li C, Liang HW, Chen JF, Yu SH (2013) Ultralight, flexible, and fire-resistant carbon nanofiber aerogels from bacterial cellulose. Angew Chem 125:2997–3001
37. Wu Q, Lü J, Qu B (2003) Preparation and characterization of microcapsulated red phosphorus and its flame-retardant mechanism in halogen-free flame retardant polyolefins. Polym Int 52:1326–1331

Chapter 5
Integrated Effect of the Triazine Based Covalent Organic Framework/-NH_2 Functionalized Black Phosphorene on Reducing Fire Hazards of Epoxy Resin Composites

5.1 Introduction

As the most thermodynamically and chemically stable allotrope of phosphorus, black phosphorus (BP), has attracted increasing interest of researchers in recent years [1, 2]. Few-layer BP (defined phosphorene) features high carrier mobility, semi-conductive properties, and a tunable intrinsic bandgap (0.3–2 eV from bulk to monolayer) [3]. These unique properties make BP a promising candidate for electrocatalysis and photocatalysis. Few-layer 2D materials are generally superior to their bulk counterparts due to their higher specific surface area and more abundant exposed active sites. As BP consists of puckered layers integrated by weak van der Waals forces, few-layer BP nanosheets can be exfoliated from bulk BP by sonication-assisted liquid exfoliation and scotch-tape solid exfoliation and sonication-assisted liquid exfoliation [4, 5]. These approaches are beneficial to control the thickness of BP nanosheets but are usually complex. Furthermore, residual organic contamination is often generated on the surface of BP. Ball milling is another widely employed technique for the exfoliation of 2D layered materials [6]. The prior work demonstrated the successful exfoliation of BP nanosheets by ball-milling with a LiOH additive, the hydroxyl functionalized and exfoliated BP nanosheets can be achieved [7]. A recent study reported the urea-assisted ball-milling synthesis of BP nanosheets, the results demonstrated that the exfoliated BP was functionalized with NH_2 groups at the edge of BP, suppressing its restacking and aiding the formation of ultrathin BP nanosheets [8]. It is desirable to exploit facile, fast, green and economical strategies to prepare phosphorene-based functional materials.

The covalent organic framework is an emerging class of two-dimensional nanomaterials with strong covalent bonds to molecular building blocks. In the stack, the two-dimensional table forms a layered overlapping structure periodically aligned with columns and functional π-electron system interactions, generating great interest in a variety of applications, such as catalysts, optoelectronic devices, energy sensors, and gas storage/separation [9–11]. So far, various covalent organic frameworks were constructed by combining covalent bonds and separating them as insoluble powders, including boronate esters [12, 13], triazines, schiff-base types, etc. [14, 15]. However, excavating the application fields of covalent organic frameworks would remain long-standing challenges. In addition, the 2D covalent organic frameworks as flame retardant additives applied in polymer composites have been scarcely investigated and reported, thereby, which is tend to be the promising application [16].

BP nanosheets, is similar to phosphorus-based flame retardants (FRs), which can act through a combination of vapor-phase and condensed-phase reactions [17]. Except for the physical barrier effect, they impart flame retardancy to polymers by promoting carbonization and char formation. Decomposition of the additive produces phosphoric acid which promotes cationic crosslinking [18]. The surface char inhibits heat feedback to promote polymer pyrolysis and the formation of fuel fragments [19]. Triazine based covalent organic framework (TOF) nanosheets are synthesized by a condensation polymerization of melamine and cyanuric chloride through a facile solvothermal process. The nanoscale 2D TOF nanosheets may exhibit a better performance in flame retardancy due to the construction of abundant nitrogen element and the high specific surface area of 2D structure. As well known, integrating two or more constituents could always display a synergistic effect and result in an extraordinary improvement in properties of polymeric composites [20]. As a novel phosphorus-nitrogen containing flame retardant system, combining the BP nanosheets and TOF nanosheets may cause the synergistic effect between these two nanomaterials.

Since BP and TOF nanosheets play roles in different combustion stages with different flame retardant action, the simultaneous integration of BP and TOF may combine the advantages of thermal stability and the flame retardant effect of these two nanomaterials. In this chapter, the functionalized BP with NH_2 groups (BP-NH_2) at the edge of BP was a synthesized by urea-assisted ball-milling process, the fabrication process of the BP-NH_2 nanosheets is show in Fig. 5.1. The triazine based covalent organic framework, using the melamine and cyanuric chloride as monomers, was synthesized via in situ condensation polymerization method, in the presence of the surface of BP-NH_2 nanosheets. The resultant BP-NH_2/TOF nanohybrids (BP-NH-TOF) with the active groups on the surface were expected to greatly decrease the interfacial tension between BP-NH-TOF and epoxy resin (EP) matrix. The effect of BP combined with the TOF nanosheets on the flame retardant and smoke toxicity suppression performances of EP were investigated.

5.2 Experimental Section

5.2.1 Raw Materials

Urea ($CO(NH_2)_2$) (99% purity), melamine ($C_3N_3(NH_2)_3$), cyanuric chloride ($C_3N_3Cl_3$), N,N-dimethylformamide (DMF), triethylamine (Et_3N) and ethanol were

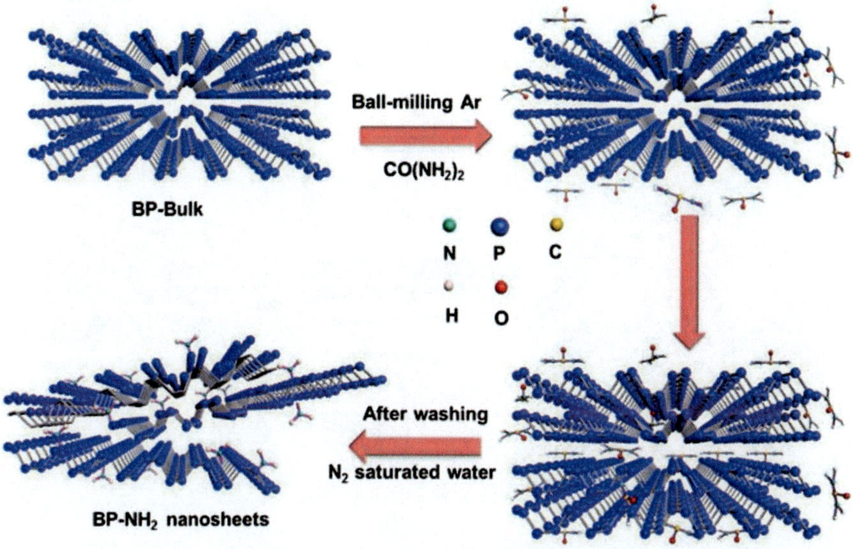

Fig. 5.1 Illustration of the fabrication process of the BP-NH_2 nanosheets

purchased from Aladdin Chemical Reagent Co. Ltd (China), and used without further purification. Suppliers of other reagents have been described in Chap. 2.

5.2.2 Preparation of BP-Bulk and BP-NH_2 Nanosheets

BP-Bulk was synthesized by a phase transformation reaction origin from RP utilizing a modified approach in Chap. 2. BP-NH_2 was synthesized by a facile solid-state mechanochemical method by ball-milling. In detail, bulk BP (200 mg) and $CO(NH_2)_2$ (400 mg, 99% purity) were mechanically ball-milled at a rate of 150 rpm for 4 h under Ar atmosphere using a planetary ball-mill machine. The milled samples were washed with N_2 saturated deionized water and ethanol several times in an Ar-filled glove box to remove $CO(NH_2)_2$. The collected powders were dispersed in absolute ethanol (200 mL) in a sealed bottle and ultrasonicated for 0.5 h in ice water. The resultant suspension was centrifuged at 3000 rpm for 30 min to remove the residual unexfoliated particles. The supernatant was collected after centrifugation at 10,000 rpm for 30 min and freeze-dried for further use.

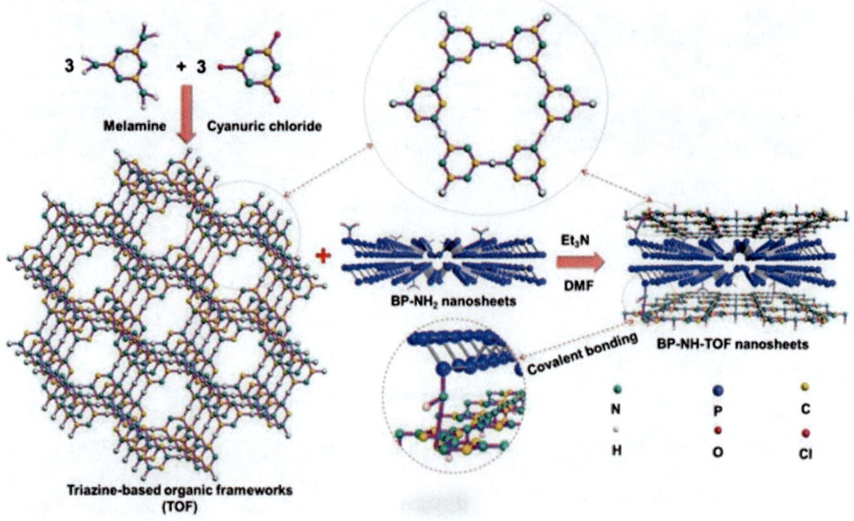

Fig. 5.2 Illustration of the fabrication process of the BP-NH-TOF nanosheets

5.2.3 Preparation of the BP-NH-TOF Nanosheets

The fabrication of BP-NH-TOF nanosheets was achieved by one pot solvothermal reaction, as shown in Fig. 5.2. In a typical procedure, the obtained BP-NH_2 nanosheets were dispersed in DMF (3.0 mg/mL, 500 mL) by ultrasonication. Melamine (0.66 g) and TEA (1.5 g) were then added to the suspension, respectively, the mixture was sonicated for 10 min. Subsequently, cyanuric chloride (0.92 g) was added and sonicated for another 10 min with stirring. The resulting mixture was transferred into a Teflon-lined stainless steel autoclave with a volume capacity of 100 mL. After sealing, the autoclave was heated to 120 °C for 24 h, and cooled naturally to room temperature. Then, the brown precipitate was isolated by centrifugation and washed with N_2 saturated deionized water and ethanol several times to remove the residuals, and then freeze-dried to collect brown powder. The as-prepared covalent-organic frameworks/BP-NH_2 nanohybrid was denoted as BP-NH-TOF. The fabrication of TOF nanosheets is similar to the fabrication process of BP-NH-TOF nanosheets without adding BP-NH_2 nanosheets.

5.2.4 Preparation of EP/BP-NH-TOF Nanocomposites

Preparation procedure of EP nanocomposites with 2.0 wt% BP-NH-TOF loading illustrated below: 0.9 g of BP-NH-TOF was dispersed in 40 mL of acetone with assistance of ultrasonication for 30 min. Follow by the corresponding 36.2 g of epoxy resin was added into the above system under sonication with mechanical stirring for 1 h. Subsequently, the redundant acetone was removed completely in a vacuum oven at 75 °C for 4 h. After that, 7.9 g of DDM was melt and poured into above blends by a rapid stirring for 1 min. Then, the resin sample was cured at 100 °C for 2 h and 150 °C for 2 h, respectively. After the curing process finished, the EP/BP-NH-TOF 2.0 sample was permitted to cool to room temperature. A similar process was utilized to prepare pure EP, EP/BP-NH-TOF0.5 (0.5 wt%), EP/BP-NH-TOF1.0 (1.0 wt%) and EP/ BP-$NH_2$2.0 (2.0 wt%) composites except the type of nanoadditives.

5.2.5 Characterization

Refer to Chaps. 2 and 3 for FTIR, XRD, Raman, XPS, TEM, SEM, AFM, TGA, TG-IR and Cone measurements.

5.3 Results and Discussion

5.3.1 Morphology and Structure Characterization of BP-NH_2 and BP-NH-TOF

To unveil the structure and composition of BP-NH_2 and BP-NH-TOF, a series of spectroscopic and morphological characterizations were employed. As shown in Fig. 5.3a, the FTIR spectrum of BP-NH_2 exists several characteristic peaks correspond to the inherent structure of BP crystal which both observed in IR spectrum of BP-Bulk, otherwise, IR spectrum of BP-NH_2 nanosheets implies the existence of a weak new absorbance at 814 cm^{-1} as compared to the BP-Bulk. This peak is assigned to the P–N stretching mode; and the apparent peaks at around 3400 cm^{-1} is attributed to the amino unit.

Figure 5.3b shows the XRD patterns of BP-Bulk and BP-NH_2 powders, the peaks are well indexed to the orthorhombic phase of BP (PDF 65-2491). Several representative diffraction peaks located at 17.1, 26.6, 34.3, 40.3, 52.5, 56.2 and 56.8° correspond to the (020), (021), (040), (002), (060), (151) and (061) interlayered planes of BP, respectively. For BP-NH_2, it shows these diffraction peaks become weaker and broader, due to the thickness of BP layers reduced after ball-milling. Besides, compared to BP-Bulk, the intensity ratio of the (111) plane and (040) plane increases in BP-NH_2 nanosheets. This phenomenon demonstrates that the crystalline plane which parallel to the layered structure is disturbed by the weakening of van der Waals interactions between the phosphorene layers [21].

Raman spectroscopy is a powerful technique used to analyze the structural integrity of the BP-Bulk and BP-NH_2 nanosheets. As shown in Fig. 5.3c, the Raman spectrum of BP-Bulk exists three representative vibrational modes located at about

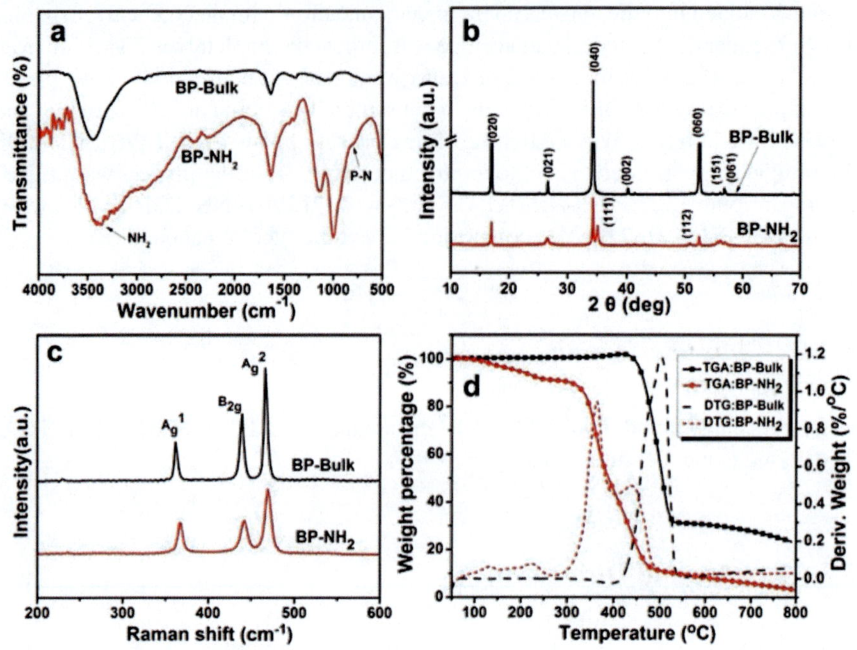

Fig. 5.3 **a** FT-IR spectra; **b** XRD patterns; **c** Raman spectra and **d** TGA curves of BP-Bulk and BP-NH_2 nanosheets

361.6, 439.7 and 466.2 cm^{-1}, which can be indexed to A_g^1, B_{2g}, and A_g^2 vibrational modes of BP, respectively. These peaks are also detected for BP-NH_2 nanosheets, revealing the retention of the vibrational structure after amino functionalization of BP. It can be seen observed that the three Raman peaks of the BP-NH_2 nanosheets blue shift slightly compared to those of the BP-Bulk. These results give the further evidence that the nanolayer numbers of BP nanosheets are reduced after -NH_2 functionalization.

Thermogravimetric analysis (TGA) was then conducted under nitrogen so as to confirm the mass loss process and NH_2 groups in BP-NH_2 nanosheets. As shown in Fig. 5.3d, bulk BP exhibits a one-step mass loss between 400 and 550 °C. Nonetheless, BP-NH_2 has two-steps mass loss until 800 °C. An obvious mass loss is observed in the first mass loss step (<250 °C) of BP-NH_2 nanosheets. This result demonstrates that the removal of the thermal elimination of the -NH_2 groups, and other defects origin from ball-milling process on the BP-NH_2 nanosheets.

X-ray photoemission spectroscopic (XPS) was carried out to investigate the valence states and chemical composition of BP-Bulk and BP-NH_2 nanosheets. In the survey spectra (Fig. 5.4), the signal peaks of C, O, N, P elements can be observed distinctly, in the spectrum of BP-NH_2 nanosheets, which demonstrates that the successful functionalization of BP-NH_2 nanosheets. As shown in Fig. 5.5a, b, the high-resolution P 2p XPS spectra of BP-Bulk sample shows two intense peaks at

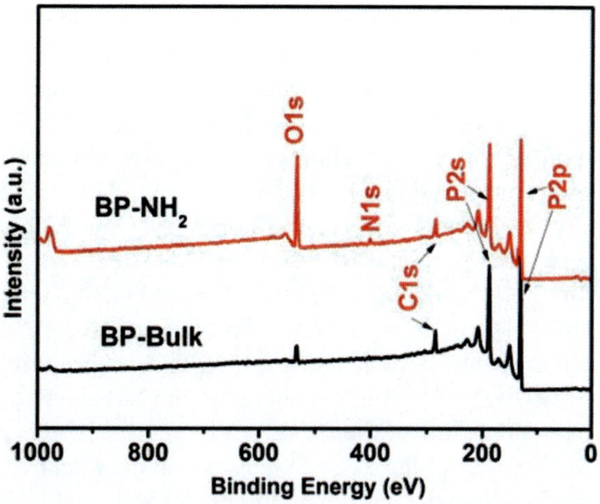

Fig. 5.4 XPS survey spectra of BP-Bulk and BP-NH_2 nanosheets

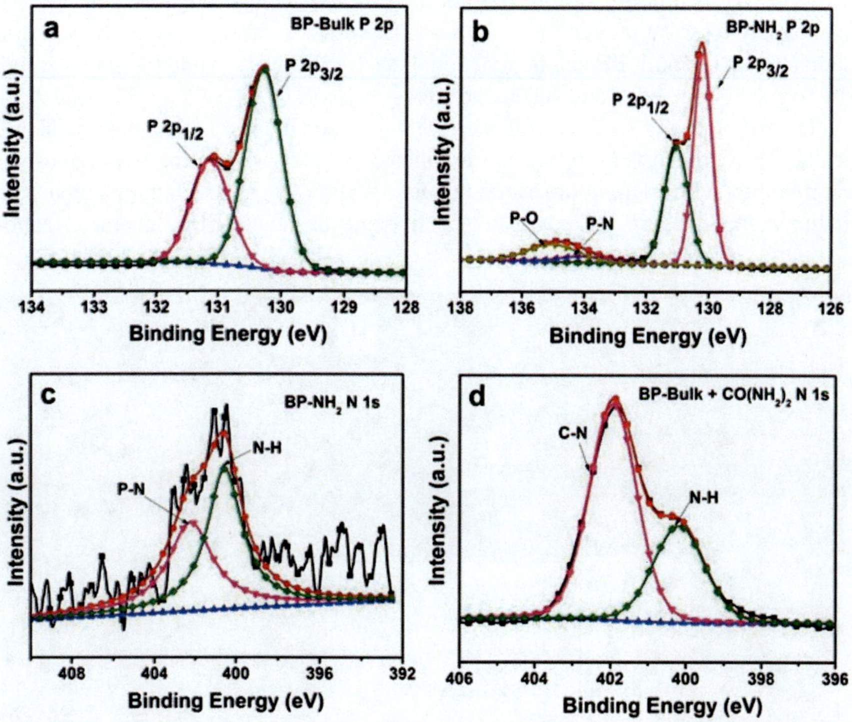

Fig. 5.5 **a**, **b** High-resolution P 2p XPS spectra of BP-Bulk and BP-NH_2; **c**, **d** high-resolution N 1s XPS spectra of BP-NH_2 and the bulk BP and $CO(NH_2)_2$ mixture

130.3 and 131.1 eV, which can be assigned to P $2p_{3/2}$ and P $2p_{1/2}$ of P-P bonds. These two peaks of BP-NH_2 nanosheets shift to lower binding energies by 0.1 eV, revealing an increase of electron concentration, a break of long-range order, and the oxidation for surface P atoms of BP nanosheets [22, 23]. BP is exceedingly inclined to oxidation in open air. However, the high-resolution P 2p XPS spectra of BP-NH_2 can be deconvoluted into four peaks at 130.2, 131.0, 133.5, 134.7 which can be attributed to P $2p_{3/2}$ and P $2p_{1/2}$ of P-P bonds, P-N and P-O bonds, respectively. In the N 1s XPS spectrum of the BP-NH_2 (Fig. 5.5c), there exist three intense peaks at 400.5 and 402.1 eV that can be ascribed to N-H and P-N bondings, respectively. For the BP and $CO(NH_2)_2$ mixture, its N 1s XPS spectrum (Fig. 5.5d) can be deconvoluted into two intense peaks at 400.2 and 401.9 eV that can be attributed to N-H and C-N bondings, respectively. The binding energy of the N–H bond of BP-NH_2 nanosheets is 0.3 eV higher than that of the BP+$CO(NH_2)_2$ mixture, this phenomenon demonstrates that the decrease of the electron density and interfacial interaction between BP and NH_2 groups. Note that a C–N bond (401.9 eV) signal is detected in the N 1s spectrum of BP+$CO(NH_2)_2$ mixture and disappears for BP-NH_2 nanosheets, revealing that the completely removal of $CO(NH_2)_2$ by washing with N_2 saturated water. XPS results indicate that the bulk BP can be functionalized by $CO(NH_2)_2$, and the -NH_2 groups are formed on the edge and surface of BP.

The BP-NH_2 and BP-NH-TOF were characterized by TEM and AFM. As shown in the TEM image (Fig. 5.6a), the exfoliated BP-NH_2 shows the well nanosheet morphology with a length of several hundred nanometers and well maintained crystallinity. In Fig. 5.6b the energy dispersive spectrometer (EDS) characterization

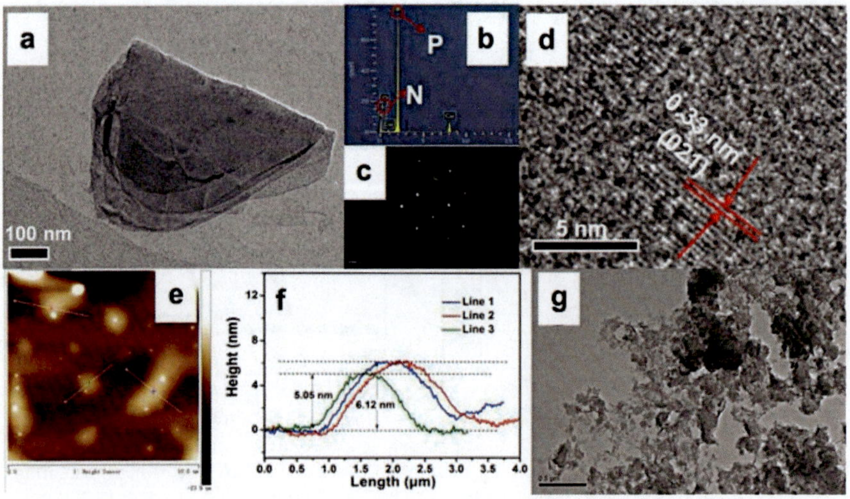

Fig. 5.6 **a** TEM image of the BP-NH_2 nanosheets and the corresponding, **b** EDS pattern, and the corresponding, **c** SAED pattern and **d** HRTEM image of BP-NH_2 nanosheets; **e** AFM image of BP-NH_2 nanosheets and **f** the corresponding height profiles taken along the lines marked in (e); **g** TEM image of the TOF nanosheets

demonstrates the elements distribution in the as-prepared BP-NH_2 nanosheets. The selected area electron diffraction (SAED) pattern (Fig. 5.6c) and the high-resolution TEM (HRTEM) image (Fig. 5.6d) show the orthorhombic crystalline structure of BP-NH_2 nanosheets. The lattice fringes of 0.33 nm is assigned to the (021) plane of BP, which is in accordance with the results in XRD analysis.

A magnified AFM image and the height profiles of few-layer BP-NH_2 nanosheets are shown in Fig. 5.6e, f. It can be observed that the BP-NH_2 nanosheets have uniform morphology with an average thickness of 6.12 nm. Normally, the thickness of single-layer phosphorene is considered as 0.53 nm. Thus, the thickness of 6.12 nm is equivalent to 11–12 layers of P atoms. As shown in Fig. 5.6g, as-prepared TOF nanosheets exhibits sheet structure up to micrometer size, similar to graphene flakes.

Additionally, the TOF nanosheets as an extra phase which grow on the BP-NH_2 nanosheets can be clearly observed in Fig. 5.7a, b. After combination of these TOF and BP-NH_2 nanosheets, the as-prepared BP-NH-TOF nanosheets maintain the well nanosheet morphology. Figure 5.7c–f shows the TEM image of BP-NH-TOF nanosheets (c) and corresponding elemental mapping images (d–f). The successful formation of BP-NH-TOF nanosheets was further verified by elemental mapping images of phosphorus (P), nitrogen (N), and carbon (C).

The FTIR spectra of TOF and BP-NH-TOF are shown in Fig. 5.8, there exist amino group at 3350 cm^{-1} which both observed in IR spectra of these two samples. Besides, some new peaks appear in the IR spectrum of BP-NH-TOF as follow, two apparent peaks at 1661 and 1536 cm^{-1} are ascribed to the triazine rings of TOF. These two peaks of BP-NH-TOF nanosheets slightly shift to higher wavenumber, revealing the interfacial interaction between TOF and BP-NH_2 nanosheets.

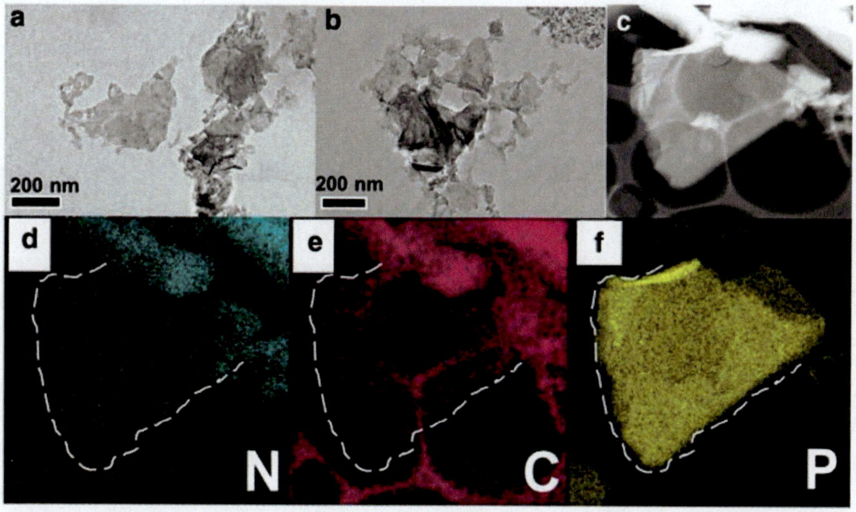

Fig. 5.7 **a**, **b** TEM image of the BP-NH-TOF nanosheets; **c** SEM image of BP-NH-TOF nanosheets and corresponding elemental mapping images of **d** nitrogen (N), **e** carbon (C) and **f** phosphorus (P)

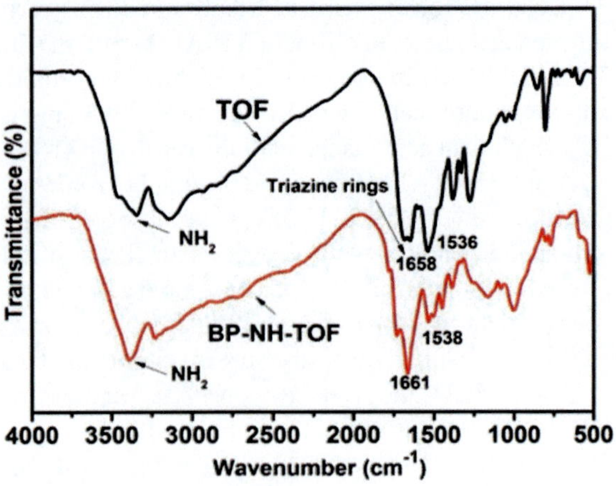

Fig. 5.8 FT-IR spectra of TOF and BP-NH-TOF nanosheets

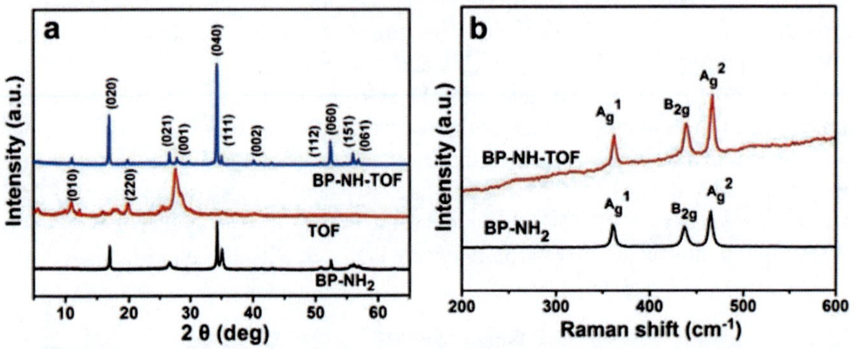

Fig. 5.9 **a** XRD patterns of BP-NH_2, TOF and BP-NH-TOF nanosheets; **b** Raman spectra of BP-NH_2 and BP-NH-TOF nanosheets

Figure 5.9a shows the XRD patterns of BP-NH_2, TOF and BP-NH-TOF powders, and the XRD pattern of BP-NH_2 as reference. Several representative diffraction peaks located at 17.1, 26.6, 34.3, 40.3, 52.5, 56.2 and 56.8° are both observed in the patterns of BP-NH_2 and BP-NH-TOF, correspond to the (020), (021), (040), (002), (060), (151) and (061) interlayered planes of BP, respectively. For TOF, it shows three broad peaks at 10.9, 20.1, and 27.5°, which are attributed to the (010), (220), and (001) facets, respectively, and the broad peak feature demonstrate the slight interlayer offsets and low crystallinity. All of these diffraction peaks in TOF appear in the pattern of BP-NH-TOF sample, this phenomenon demonstrates that the success formation of TOF nanosheets attach on the surface of BP-NH_2 nanosheets.

The structural integrity of the BP-NH-TOF was further confirmed by analyzing the Raman spectra (Fig. 5.9b). Three representative vibrational modes of BP-NH_2 are assigned to the peaks of A_g^1 at 361.9 cm^{-1}, B_{2g} at 439.2 cm^{-1}, and A_g^2 at 467.6 cm^{-1}.

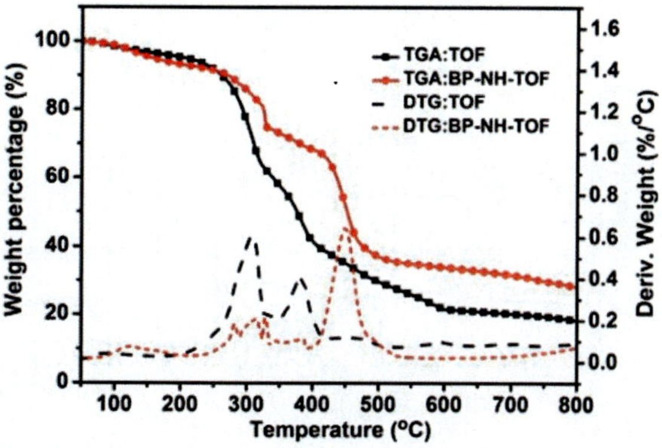

Fig. 5.10 TGA and DTG curves of TOF and BP-NH-TOF nanosheets

These peaks are also detected for BP-NH-TOF sample, demonstrating the retention of the vibrational structure after combination of TOF and $BP\text{-}NH_2$ nanosheets.

Thermogravimetric analysis (TGA) and DTG curves of TOF and BP-NH-TOF nanosheets are shown in Fig. 5.10. Pure TOF nanosheets exhibit a two-step mass loss between 200 and 350 and 350–500 °C, respectively. However, BP-NH-TOF has three steps mass loss until 800 °C. The last mass loss step which different from TOF is assigned to the main degradation of BP. These results give the further evidence that the success formation of BP-NH-TOF dual nanosheets.

In the XPS survey spectra of $BP\text{-}NH_2$, TOF and BP-NH-TOF nanosheets (Fig. 5.11a), the survey spectrum of $BP\text{-}NH_2$ shows that the sample includes O, C, P and N elements. The surface of TOF was composed of C, O and N elements. As expected, the signal peaks of C, O, N, P elements can be observed distinctly, in the spectrum of BP-NH-TOF nanosheets, and the elements content of N is increased. Figure 5.11b shows the high-resolution C 1s spectrum of TOF nanosheets, is deconvoluted into two peaks at 288.1 and 285.0 eV originated from sp2-hybridized carbon (NC=N) and adventitious carbon, respectively [24]. Besides, the high-resolution spectrum of N 1s is also deconvoluted into two peaks located at 398.7 and 399.6 eV, respectively, in Fig. 5.11c. The peak at higher binding energies could be assigned to nitrogen species in the forms of C-NH-C; and the lower binding energies could be ascribed to sp2 hybridized nitrogen (C-N=C) [25]. Compared to TOF, the C 1s signal of BP-NH-TOF in Fig. 5.11d is also divided into two peaks at 288.1 and 285.0 eV, correspond to N-C=N and C-C bonds, respectively. As shown in Fig. 5.11e, In the N1s XPS spectrum of the BP-NH-TOF, there exist three intense peaks at 400.3 and 399.2 eV that can be ascribed to C-NH-C and C-N=C bondings, respectively, these bonds of BP-NH-TOF nanosheets is slightly shift to higher binding energy than pure TOF, this phenomenon demonstrates that the formation of interfacial interaction between TOF and $BP\text{-}NH_2$ nanosheets. For P 2p peaks of BP-NH-TOF in Fig. 5.11f, which can be deconvoluted into four peaks at 129.6, 130.5, 133.4, and 134.4 eV correspond to P $2p_{3/2}$ and P $2p_{1/2}$ of P-P bonds, P-N and P-O bonds, respectively.

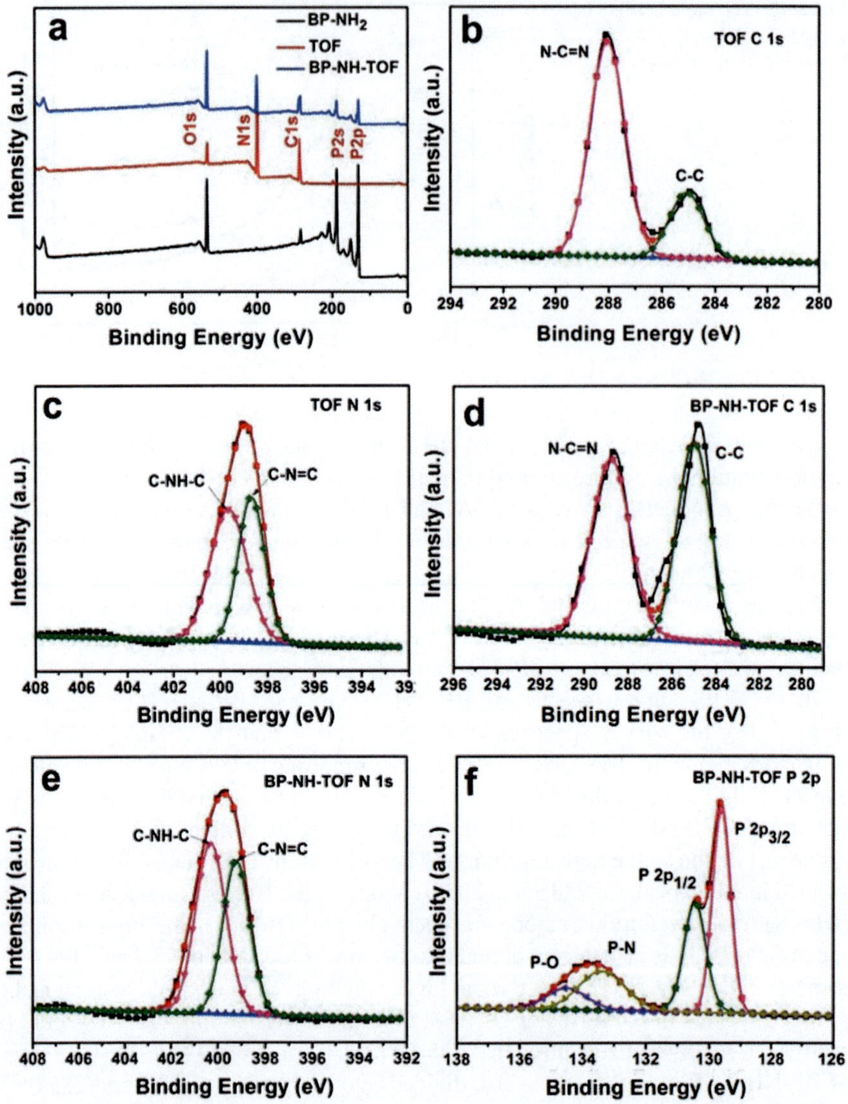

Fig. 5.11 **a** XPS survey spectra of BP-NH_2, TOF and BP-NH-TOF; High-resolution, **b** C 1s and, **c** N 1s XPS spectra of TOF nanosheets; High-resolution **d** C 1s, **e** N 1s and **f** P 2p XPS spectra of BP-NH-TOF

5.3.2 *Morphology and Structure Characterization of EP/BP-NH-TOF Composites*

Figure 5.12 shows the SEM images of freeze-fractured surface for EP, EP/BP-$NH_2$2.0 and EP/BP-NH-TOF2.0 nanocomposites. In the SEM image of pure EP (Fig. 5.12a),

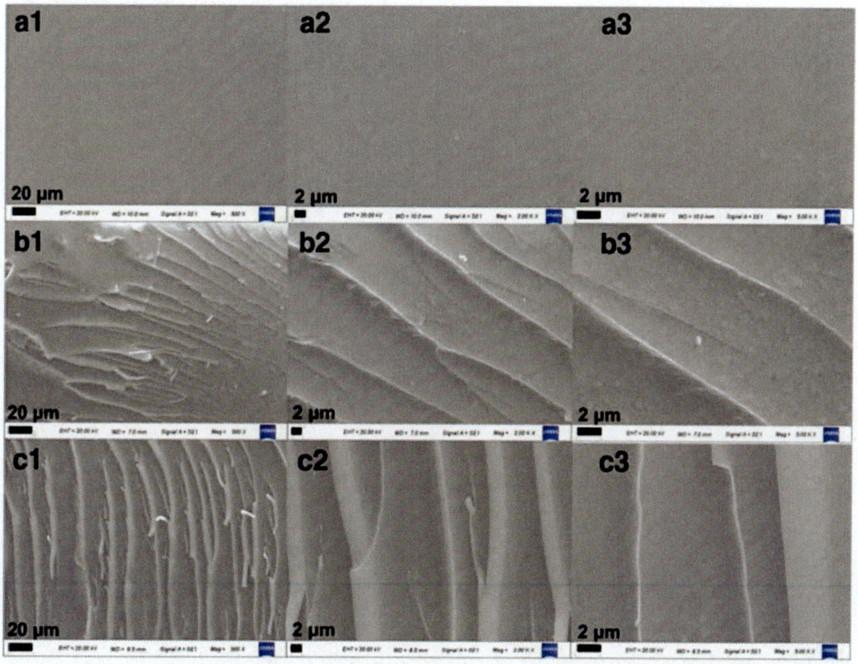

Fig. 5.12 SEM images of the **a** pure EP, **b** EP/BP-$NH_2$2.0 and **c** EP/BP-NH-TOF2.0 nanocomposites in different magnification

the fractured surface of this sample is extremely smooth and shows a no-crinkled morphology. As shown in Fig. 5.12b, it can be observed that the fracture surface of EP/BP-$NH_2$2.0 is relative rough with small-crinkled morphology, indicating the formation of interface interaction between BP-$NH_2$2.0 nanosheets and EP matrix. For the EP/BP-NH-TOF2.0 nanocomposite (Fig. 5.12c), it exhibits a more rough fractured surface with large-crinkled morphology, compared to those of pure EP and EP/BP-$NH_2$2.0. This may be due to the interpretations that combination of BP-NH_2 with TOF nanosheets leading to the enhanced interfacial adhesion between BP-NH-TOF and EP matrix [26].

Figure 5.13a, b show the FTIR spectra and XRD patterns of the EP nanocomposites. The FTIR spectra of the pure EP, EP/BP-NH_2 and EP/BP-NH-TOF composites are shown in Fig. 5.13a. Pure EP exhibits several typical absorption peaks, and these typical absorption bands are observed in the FTIR spectra of the EP/BP-NH_2 and EP/BP-NH-TOF composites as well. The XRD patterns of pure EP, EP/BP-NH_2 and EP/BP-NH-TOF composites are presented in Fig. 5.13b. Compared to pure EP, There exists several sharp diffraction peaks in the EP/BP-NH_2 and EP/BP-NH-TOF composites are assigned to (020), (021), (040), (111), (112), (060), (151) and (061) interlayered planes of BP. Besides, in the XRD patterns of EP/BP-NH-TOF, the weak peaks at $2\theta = 10.9$ and 20.0° are ascribed to the (010) and (220) plane of crystalline structure of TOF.

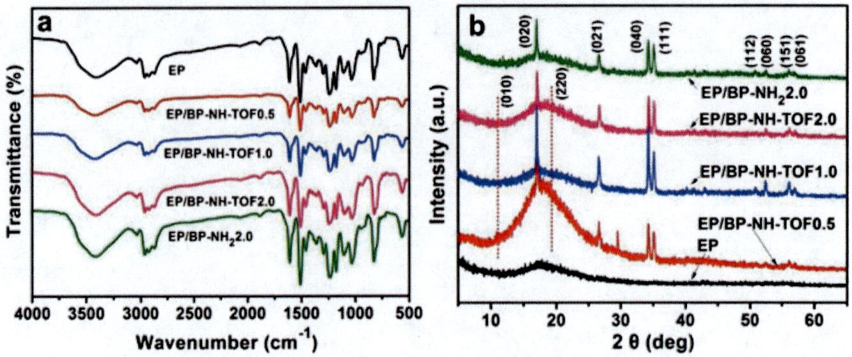

Fig. 5.13 **a** FTIR spectra and **b** XRD patterns of EP and EP/BP-NH-TOF nanocomposites

5.3.3 *Thermal and Mechanical Properties of EP/BP-NH-TOF Composites*

The effect of BP-NH-TOF nanosheets on the thermal property of EP nanocomposites was studied by TGA analysis under nitrogen (N_2). As shown in Fig. 5.14a, the pure EP shows a one-step mass loss in the region of 300–500 °C, and EP/BP-NH_2 and EP/BP-NH-TOF nanocomposites exhibit similar pyrolysis behavior to pure EP. The onset degradation temperature of the EP/BP-NH-TOF nanocomposites changed slightly, however, with increasing of the EP/BP-NH-TOF content, their char residues at 800 °C gradually increased significantly, higher than that of pure EP, owing to the physical barrier and catalytic charring effect of the BP-NH-TOF nanosheets. Particularly, the initial decomposition temperature of EP/BP-$NH_2$2.0 sample is near to pure EP, with a char yield of 15.6% at 800 °C, lower than char yield (17.5%) of EP/BP-NH-TOF2.0 nanocomposites with the same loading, due to the earlier decomposition of EP matrix resulting from the catalytic charring effect of BP.

In the derivative thermogravimetric analysis (DTG) curves of EP nanocomposites (Fig. 5.14b), the maximum mass loss rates of EP/BP-NH_2 and EP/BP-NH-TOF composites are much lower than that of pure EP. BP-NH-TOF nanosheets play as a physical barrier which can inhibit oxygen exchange and diffusion of pyrolysis products, the well-organized P-N system which promote the char formation, thus improving thermal resistance of the EP nanocomposites against degradation.

The mechanical and thermal properties were studied by DMA to verify these performances were improved after introducing EP/BP-NH_2 and EP/BP-NH-TOF nanosheets into EP matrix. The storage modulus and tan δ of the EP/BP-NH_2 and EP/BP-NH-TOF nanocomposites as a function of temperature are shown in Fig. 5.15a, b. The pure EP exhibits a storage modulus at room temperature of approximately 21,115.59 MPa. With increasing the content of BP-NH-TOF nanosheets from 0.5 to 2.0 wt%, the storage modulus values at room temperature of EP/BP-NH-TOF nanocomposites are significantly improved by 51.4%, 67.1%, 14.4%, respectively,

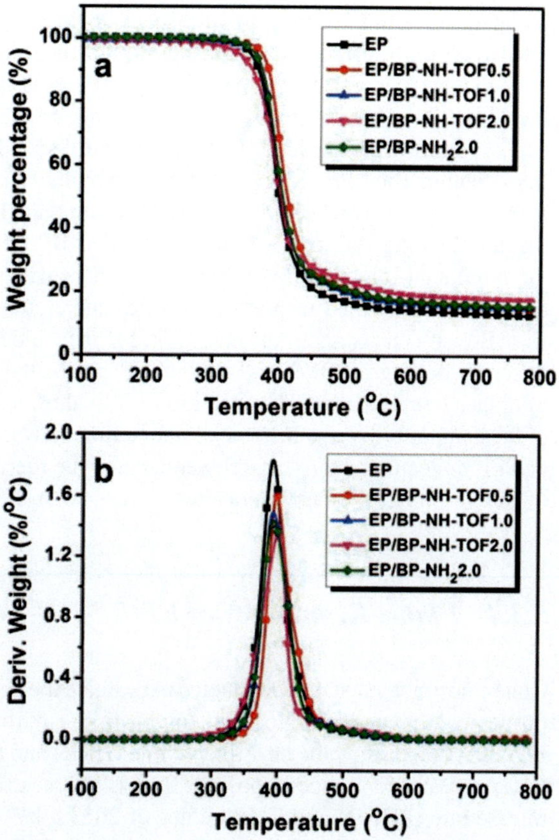

Fig. 5.14 **a** TGA curves and, **b** DTG curves of the pure EP and its nanocomposites with different BP-NH-TOF contents

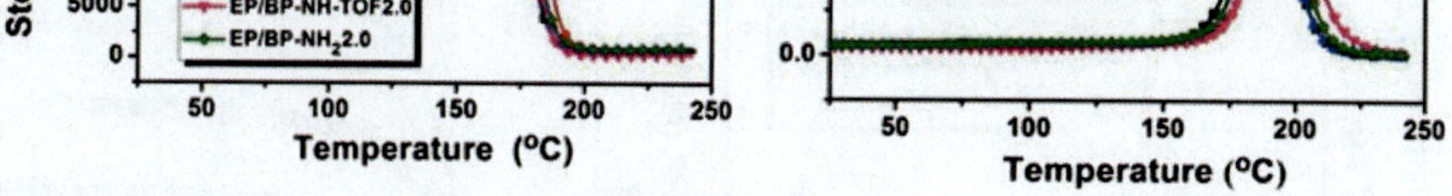

Fig. 5.15 **a** Storage modulus (E') curves and **b** Tan δ curves of the pure EP and its nanocomposites with different BP-NH-TOF contents as a function of temperature

compared to pure EP. These enhancements in the storage modulus are directly correlated to the high stiffness of BP-NH-TOF nanosheets and the chemical bonding of the TOF to epoxy network. Compared to EP/BP-NH-TOF0.5 and EP/BP-NH-TOF1.0, the EP/BP-NH-TOF2.0 exhibits lower storage modulus, due to the re-aggregation and non-uniform dispersion of BP-NH-TOF fillers in EP matrices with high loading. In particular, The EP/BP-$NH_2$2.0 sample exhibits a storage modulus at room temperature of 22,323.16 MPa, which is relative lower than that of EP/BP-NH-TOF2.0 with the same loading, due to the existence of TOF in BP-NH-TOF nanosheets can improve the interfacial interaction between these fillers with EP matrix. The peak value of tan δ can be ascribed to glass transition temperature (T_g) of the EP/BP-NH_2 and EP/BP-NH-TOF nanocomposites (Fig. 5.15b). In detail, with the incorporating of different loadings of BP-NH-TOF into EP matrix, the peaks of tan δ are slightly shift to higher temperatures. The increase of T_g values is attributed to the enhancement of the compatibility and interfacial interaction between the BP-NH-TOF dual-sheets and EP molecular chains, thereby enhancing the thermal and mechanical peoperties of EP/BP-NH-TOF nanocomposites.

5.3.4 Flame Retardancy of EP/BP-NH-TOF Composites

Cone calorimeter test was conducted to evaluate the fire performance of EP nanocomposites which can imitate real combustion environment of polymer materials. Figure 5.16a, b show the heat release rate (HRR) and total heat release (THR) versus time curves of EP nanocomposites. Pure EP burns dramatically with high peak heat release rate (PHRR) and THR values of 2053.4 kW/m^2 and 160.7 MJ/m^2, respectively. For EP/BP-NH-TOF nanocomposites, the PHRR values of the these nanocomposites decreased by 16.7–60.1%; and the THR values of the these nanocomposites decreased by 17.5–43.4%, with increasing BP-NH-TOF content from 0.5 to 2.0 wt%. Particularly, the addition of 2 wt% BP-NH-TOF in EP resulting in a maximum 60.1%

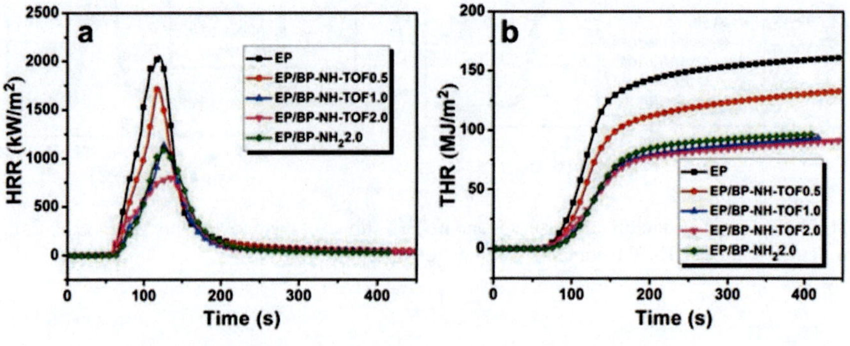

Fig. 5.16 **a** HRR and **b** THR versus time curves of the pure EP and its nanocomposites with different BP-NH-TOF contents

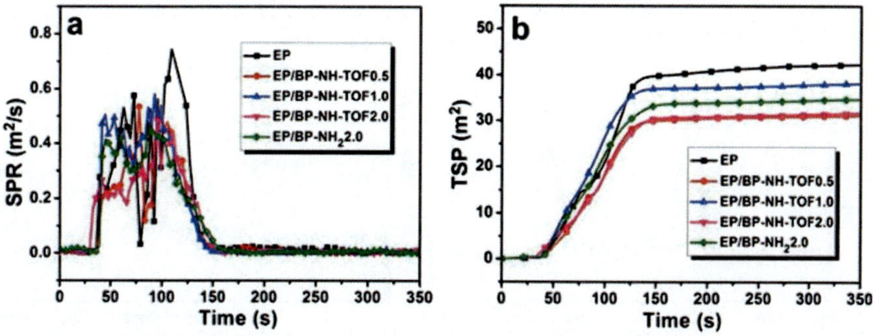

Fig. 5.17 **a** SPR and **b** TSR versus time curves of the pure EP and its nanocomposites with different BP-NH-TOF contents from cone tests

decrease in PHRR, a maximum 43.4% decrease in THR during combustion, demonstrating the improved flame retardant properties of EP/BP-NH-TOF nanocomposites with the increasing of BP-NH-TOF loading. The PHRR and THR values for the EP/BP-NH_2 sample are decreased by 46.5% and 39.6%, respectively, the reduction scale are lower than those of EP/BP-NH-TOF2.0 nanocomposites with the same loading.

The smoke production rate (SPR) and total smoke release (TSR) versus time curves of EP nanocomposites are shown in Fig. 5.17a, b. Pure EP can easily release heavy smoke during combustion with higher SPR and TSR values due to its multi-aromatic structure. As expected, the *SPR* and TSR values of the EP/BP-NH-TOF nanocomposites with various loadings are obviously decreased compared to those of pure EP, and EP/BP-NH-TOF2.0 shows the lower *SPR* and TSR values, than those of EP/BP-NH_2. The significant reductions in the fire hazards of EP composites are attributed to the physical barrier effect of BP-NH-TOF dual-sheets; and these dual-sheets building a well-organized P-N system with cooperative catalytic charring effect which promote the char formation, thereby retarding the diffusion of pyrolysis products and forbid the inner unburned materials exposed to fire during combustion.

To further study the thermal degradation behavior of EP/BP-$NH_2$2.0 and EP/BP-NH-TOF2.0 nanocomposites, the pyrolysis products released from EP, EP/BP-$NH_2$2.0 and EP/BP-NH-TOF2.0 nanocomposites were evaluted by TG-FTIR technique. The intensities of the several typical volatile products versus time of EP nanocomposites are shown in Fig. 5.18. Several characteristic peaks for the evolved pyrolysis products of EP nanocomposites as follow: the peak at 2930 cm^{-1} is assigned to absorption of hydrocarbons; the peak at 2360 cm^{-1} corresponds to absorption of carbon dioxide (CO_2); the peak at 2190 cm^{-1} is ascribed to absorption of carbon monoxide (CO); the peak at 1740 cm^{-1} is attributed to absorption of carbonyl compounds and the peak at 1510 cm^{-1} is assigned to absorption of aromatic compounds [27, 28]. As shown in Fig. 5.18a–f, the maximum absorbance intensities of characteristic pyrolysis products for EP/BP-$NH_2$2.0 and EP/BP-NH-TOF2.0 samples are significantly reduced, lower than those of pure EP, including CO and

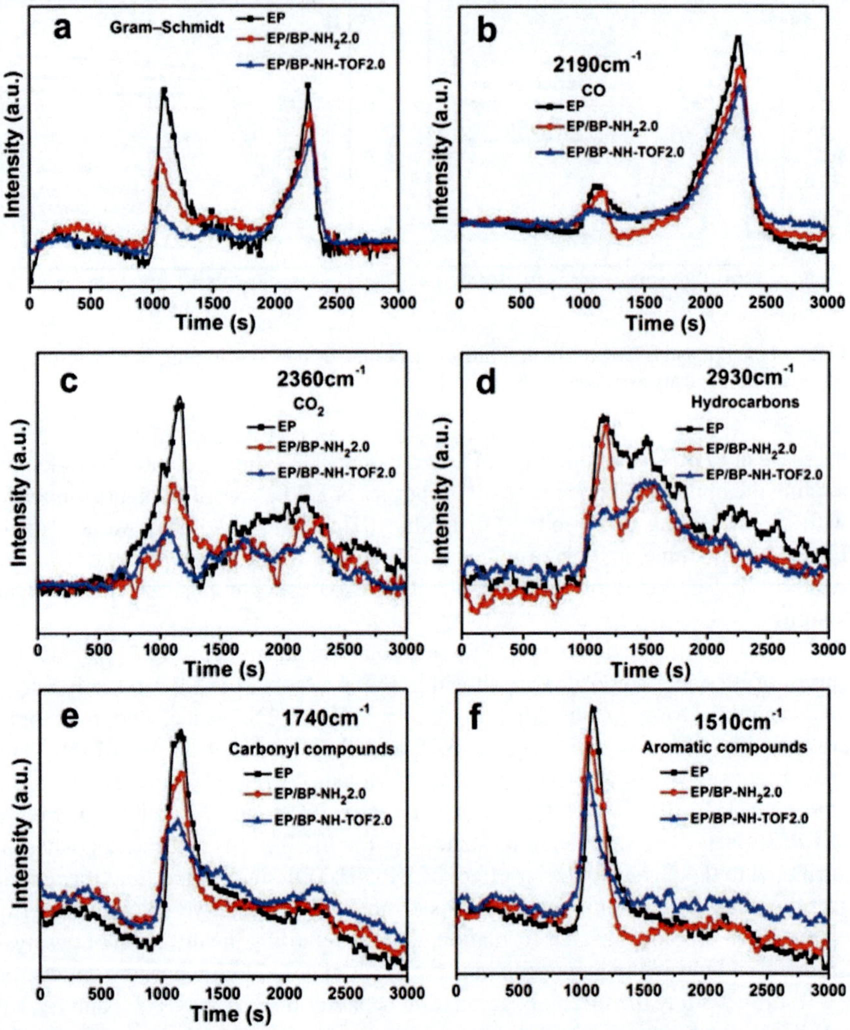

Fig. 5.18 Absorbance of pyrolysis products for EP, EP/BP-$NH_2$2.0 and EP/BP-NH-TOF2.0 sample versus time: **a** total pyrolysis products; **b** CO; **c** CO_2; **d** hydrocarbons; **e** carbonyl compounds and **f** aromatic compounds

CO_2, hydrocarbons, aromatic compounds and carbonyl compounds. In particular, EP/BP-NH-TOF2.0 nanocomposite shows lowest absorbance intensity of typical volatile products among pure EP and EP/BP-$NH_2$2.0 nanocomposites. The reduction of intensities of flammable pyrolysis products including hydrocarbons, carbonyl and aromatic compounds is beneficial to decreasing smoke and heat release; CO is considered to be the primary toxic substance in the combustion process of EP, and the

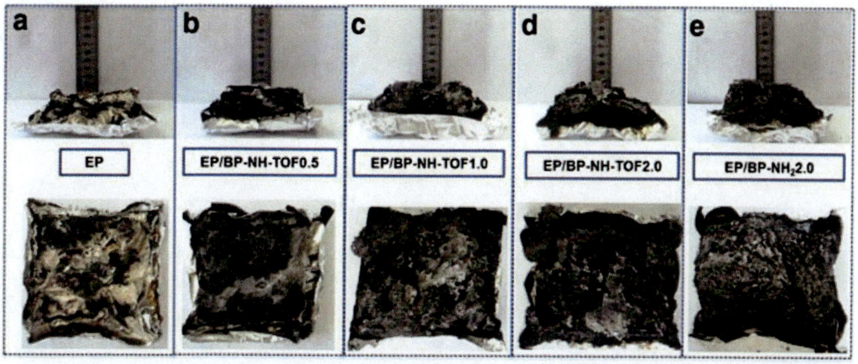

Fig. 5.19 Digital photos of the external residues from top view and side view for EP and its nanocomposites

decrease of CO intensity contributes to decrease smoke toxicity, thereby improving the fire safety.

5.3.5 Condensed Phase Analysis of EP/BP-NH-TOF Composites

In order to explore reasonable flame-retardant mechanism, the char residues of EP, EP/BP-NH_2 and EP/BP-NH-TOF nanocomposites origined from cone test were investigated. The digital photos of the external residues from top view and side view for these EP nanocomposites are shown in Fig. 5.19. A small amount of cracked char remains in pure EP sample (Fig. 5.19a). As a contrast, with the increasing the loadings of BP-NH-TOF from 0.5 to 2.0 wt% (Fig. 5.19b–d), the size and amount of char residues are gradually increased. The digital photo of EP/BP-NH-TOF2.0 nanocomposite shows an expanded and continuous dense char layer (Fig. 5.19d). The EP/BP-$NH_2$2.0 nanocomposite also exhibits an expanded char layer with small holes on the surface (Fig. 5.19e). The quality of char layer for EP/BP-NH-TOF2.0 is better than that of EP/BP-$NH_2$2.0 sample with the same loading.

Figure 5.20 shows the microstructures of external (5.20a–c) and interior (11d–f) char residues for EP, EP/BP-NH_2 and EP/BP-NH-TOF nanocomposites. As shown in the SEM image of external char layer for pure EP (Fig. 5.20a), this sample exhibits a flaky morphology with big opening hole on the surface. As a contrast, the external residue of EP/BP-$NH_2$2.0 sample shows a continuous and compact surface with no holes (Fig. 5.20b). However, after the EP/BP-NH-TOF2.0 nanocomposite combustion from cone, it can be seen from Fig. 5.20c that a continuous and more compact char layer is formed. Figure 5.20d–f show the SEM images of internal residues which further confirm that the internal char residue of EP/BP-NH-TOF sample is more compact than those of pure EP and EP/BP-NH_2 sample. It can be observed that

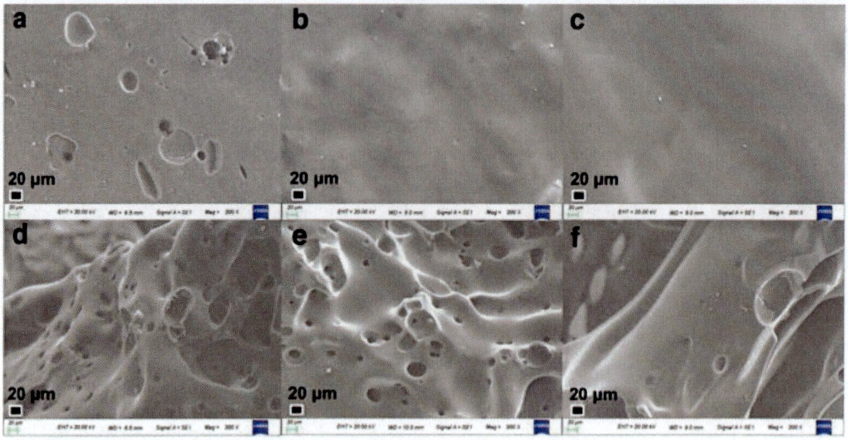

Fig. 5.20 SEM images of external **a–c** and interior, **d–f** char residues for EP, EP/BP-NH_2 and EP/BP-NH-TOF2.0 nanocomposites

many big opening holes on the surface of internal residues for pure EP and EP/BP-NH_2 sample, as compared, the internal char of EP/BP-NH-TOF sample exists a more condense inwall with few holes, theses phenomenon demonstrates that the physical barrier effect and catalytic charring effect of BP-NH-TOF nanosheets can improve the quality of char layer during the combustion. As well known, the char layer with continuous and denser surface is beneficial to retarding the mass and heat transfer, and release of pyrolysis products, thereby enhancing the fire safety performance.

Raman spectroscopy was used to evaluate the composition and structure of the char residue. As shown in Fig. 5.21a–c, it can be observed that two representative peaks in the Raman spectra of EP, EP/BP-$NH_2$2.0 and EP/BP-NH-TOF2.0 nanocomposites located at 1596 and 1365 cm^{-1}, are defined as G peak and D peak, respectively. In previous works, the area ratio of D peak to G peak (I_D/I_G) represents the graphitization degree of char residue, and higher I_D/I_G value equals to lower graphitization degree [29, 30]. The carbonized char layer with high graphitization degree is beneficial to preventing the internal material ignition. Particularly, the I_D/I_G value of pure EP is 3.23, whereas the EP/BP-$NH_2$2.0 nanocomposite shows a I_D/I_G value of 3.05, the EP/BP-NH-TOF2.0 sample exhibits lower I_D/I_G value (2.71) among three samples, implying the higher graphitization degree, owing to the catalytic charring effect of BP-NH-TOF dual-sheets during EP combustion.

XRD patterns of the char residues for EP and EP/BP-NH-TOF nanocomposites after cone tests are shown in Fig. 5.22a. In the XRD pattern of EP and its nanocomposites, there exist a wide diffraction peak at 23°, assigned to (002) diffraction of graphite, revealing the formation of graphitized carbon [31], however, there is no peak assigned to BP crystal, indicating most BP-NH_2 and BP-NH-TOF are completely degraded after high temperature combustion. XPS analysis provides more detail information on the composition and structure of the residual chars. The XPS survey spectra of EP and EP/BP-NH-TOF nanocomposites are shown in Fig. 5.22b, the

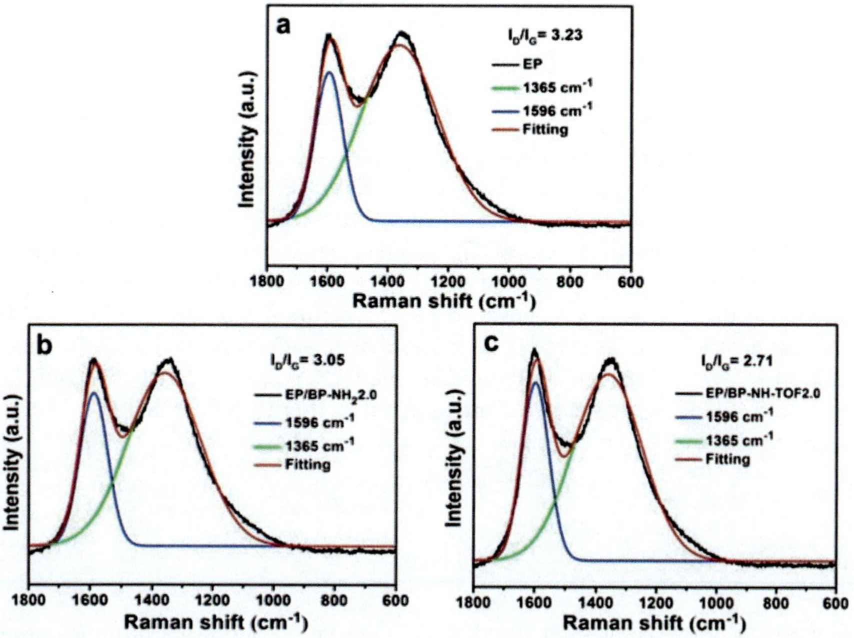

Fig. 5.21 Raman spectra of the char residues of **a** EP, **b** EP/BP-NH_2 and **c** EP/BP-NH-TOF2.0 nanocomposites

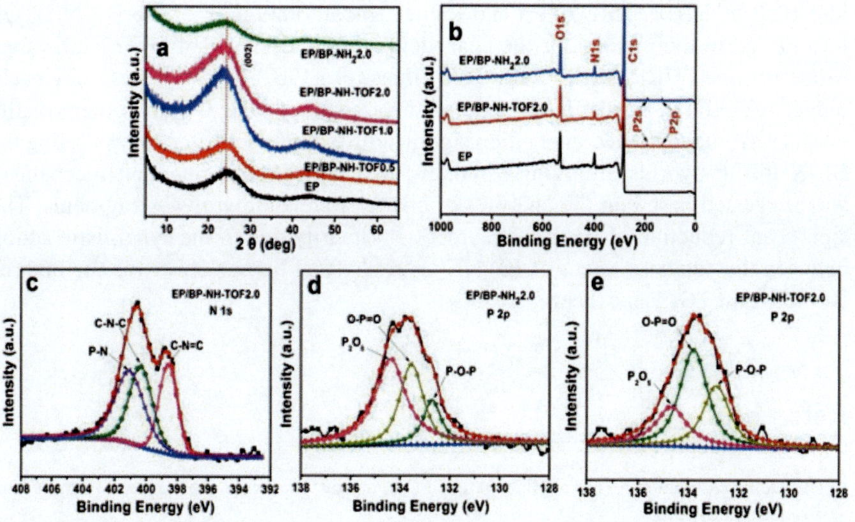

Fig. 5.22 **a** XRD patterns of the char residues for EP nanocomposites; **b** XPS survey spectra of the residual char for EP nanocomposites; **c** High-resolution N 1s XPS spectra of the residues for EP/BP-NH-TOF2.0 sample; P 2p XPS spectra of the residues for **d** EP/BP-$NH_2$2.0 and **e** EP/BP-NH-TOF2.0 composites

surface of pure EP sample is consist of C, N and O elements, the extra P element appear in the residual char of EP/BP-NH_2 and EP/BP-NH-TOF nanocomposites, because of the PO_x and its derivate from the decomposition of BP remain in the char residue. Figure 5.22c shows the high-resolution XPS spectra of EP/BP-NH-TOF in the N 1s regions, there exists three intense peaks can be assigned to P-N, C-N-C and C-N=C bondings, respectively, implying the formation of crosslinked phosphorus oxynitride. Besides, it is demonstrate that a part of TOF nanosheets remain in the char residue of EP nanocomposites after combustion. These results are also proved by high-resolution P 2p XPS spectrum of EP/BP-NH_2 and EP/BP-NH-TOF nanocomposites in Fig. 5.22d, e. As shown in Fig. 5.22d, the P 2p peaks of EP/BP-NH_2 can be deconvoluted into three peaks correspond to P-O-P, O-P=O and P_2O_5 bonds, respectively. As compared, in Fig. 5.22e, the P 2p peaks of EP/BP-NH-TOF can be deconvoluted into three peaks are ascribed to P-O-P, O-P=O and P_2O_5 bonds, respectively, revealing the formation of crosslinked phosphorus oxynitride [32].

5.4 Summary

In this work, the -NH_2 functionalized BP (BP-NH_2) was synthesized via urea-assisted ball-milling procedure. The triazine based covalent organic framework/BP-NH_2 (BP-NH-TOF) nanohybrid was synthesized via in situ condensation polymerization method using BP-NH_2 as a template, and its composition and structure was identified by FT-IR, XRD, XPS and Raman. Incorporation of 2 wt% BP-NH-TOF into EP led to the increase of the char yield and the decrease of DTG peak value. Moreover, the PHRR, THR, TSR, SPR values for EP/BP-NH-TOF were obviously reduced. TG-FTIR results presented that the amount of toxic CO and other volatile products from the EP decomposition significantly suppressed after incorporating the BP-NH-TOF hybrids, implying a reduced toxicity. A flame-retardant mechanism was suggested based on the analyses of char residue and pyrolysis fragments. The significant reduction of the fire hazard was primarily due to the synergistic action between the catalytic effect of BP-NH_2 and physical barrier effect for the both of BP-NH_2 and TOF nanosheets.

References

1. Castellanos-Gomez A (2015) Black phosphorus: narrow gap, wide applications. J Phys Chem Lett 6:4280–4291
2. Carvalho A, Wang M, Zhu X, Rodin AS, Su H, Neto AHC (2016) Phosphorene: from theory to applications. Nat Rev Mater 1:16061
3. Gong P-L, Liu D-Y, Yang K-S, Xiang Z-J, Chen X-H, Zeng Z, Shen S-Q, Zou L-J (2016) Hydrostatic pressure induced three-dimensional Dirac semimetal in black phosphorus. Phys Rev B 93:195434

4. Lin S, Chui Y, Li Y, Lau SP (2017) Liquid-phase exfoliation of black phosphorus and its applications. FlatChem 2:15–37
5. Gibaja C, Rodriguez-San-Miguel D, Ares P, Gómez-Herrero J, Varela M, Gillen R, Maultzsch J, Hauke F, Hirsch A, Abellán G (2016) Few-layer antimonene by liquid-phase exfoliation. Angew Chem Int Ed 55:14345–14349
6. León V, Rodriguez AM, Prieto P, Prato M, Vázquez E (2014) Exfoliation of graphite with triazine derivatives under ball-milling conditions: preparation of few-layer graphene via selective noncovalent interactions. ACS Nano 8:563–571
7. Zhu X, Zhang T, Sun Z, Chen H, Guan J, Chen X, Ji H, Du P, Yang S (2017) Black phosphorus revisited: a missing metal-free elemental photocatalyst for visible light hydrogen evolution. Adv Mater 29:1605776
8. Shao L, Sun H, Miao L, Chen X, Han M, Sun J, Liu S, Li L, Cheng F, Chen J (2018) Facile preparation of NH_2-functionalized black phosphorene for the electrocatalytic hydrogen evolution reaction. J Mater Chem A 6:2494–2499
9. Cote AP, Benin AI, Ockwig NW, O'keeffe M, Matzger AJ, Yaghi OM (2005) Porous, crystalline, covalent organic frameworks. Science 310:1166–1170
10. Feng X, Ding X, Jiang D (2012) Covalent organic frameworks. Chem Soc Rev 41:6010–6022
11. Ding S-Y, Wang W (2013) Covalent organic frameworks (COFs): from design to applications. Chem Soc Rev 42:548–568
12. Zwaneveld NA, Pawlak R, Abel M, Catalin D, Gigmes D, Bertin D, Porte L (2008) Organized formation of 2D extended covalent organic frameworks at surfaces. J Am Chem Soc 130:6678–6679
13. Spitler EL, Giovino MR, White SL, Dichtel WR (2011) A mechanistic study of Lewis acid-catalyzed covalent organic framework formation. Chem Sci 2:1588–1593
14. Gomes R, Bhanja P, Bhaumik A (2015) A triazine-based covalent organic polymer for efficient CO_2 adsorption. Chem Commun 51:10050–10053
15. Wen P, Zhang C, Yang Z, Dong R, Wang D, Fan M, Wang J (2017) Triazine-based covalent-organic frameworks: a novel lubricant additive with excellent tribological performances. Tribol Int 111:57–65
16. Mu X, Zhan J, Feng X, Yuan B, Qiu S, Song L, Hu Y (2017) Novel melamine/o-phthalaldehyde covalent organic frameworks nanosheets: enhancement flame retardant and mechanical performances of thermoplastic polyurethanes. ACS Appl Mater Interfaces 9:23017–23026
17. Van der Veen I, de Boer J (2012) Phosphorus flame retardants: properties, production, environmental occurrence, toxicity and analysis. Chemosphere 88:1119–1153
18. Zhao C-X, Liu Y, Wang D-Y, Wang D-L, Wang Y-Z (2008) Synergistic effect of ammonium polyphosphate and layered double hydroxide on flame retardant properties of poly (vinyl alcohol). Polym Degrad Stabil 93:1323–1331
19. Ball R, McIntosh A, Brindley J (1999) The role of char-forming processes in the thermal decomposition of cellulose. Phys Chem Chem Phys 1:5035–5043
20. Zhao Y-H, Zhang Y-F, Wu Z-K, Bai S-L (2016) Synergic enhancement of thermal properties of polymer composites by graphene foam and carbon black. Compos B Eng 84:52–58
21. Cai Y, Ke Q, Zhang G, Zhang Y-W (2015) Energetics, charge transfer, and magnetism of small molecules physisorbed on phosphorene. J Phys Chem C 119:3102–3110
22. Zhao Y, Chen Y, Zhang Y-H, Liu S-F (2017) Recent advance in black phosphorus: properties and applications. Mater Chem Phys 189:215–229
23. Ryder CR, Wood JD, Wells SA, Hersam MC (2016) Chemically tailoring semiconducting two-dimensional transition metal dichalcogenides and black phosphorus. ACS Nano 10:3900–3917
24. Shu T, Yang W, Li K, Yan L, Dai Y, Guo H (2015) Design of silver-deposited carbon nitride nanotubes by a one-step solvothermal treatment strategy and their efficient visible-light photocatalytic activity toward methyl orange degradation. Energ Environ Focus 4:107–115
25. Kaswan P, Porter A, Pericherla K, Simone M, Peters S, Kumar A, DeBoef B (2015) Oxidative cross-coupling of sp3-and sp2-Hybridized C-H bonds: vanadium-catalyzed aminomethylation of imidazo [1, 2-a] pyridines. Org Lett 17:5208–5211

26. Yang L, Phua SL, Teo JKH, Toh CL, Lau SK, Ma J, Lu X (2011) A biomimetic approach to enhancing interfacial interactions: polydopamine-coated clay as reinforcement for epoxy resin. ACS Appl Mater Interfaces 3:3026–3032
27. Qian X, Song L, Yu B, Wang B, Yuan B, Shi Y, Hu Y, Yuen RK (2013) Novel organic–inorganic flame retardants containing exfoliated graphene: preparation and their performance on the flame retardancy of epoxy resins. J Mater Chem A 1:6822–6830
28. Yu B, Wang X, Qian X, Xing W, Yang H, Ma L, Lin Y, Jiang S, Song L, Hu Y (2014) Functionalized graphene oxide/phosphoramide oligomer hybrids flame retardant prepared via in situ polymerization for improving the fire safety of polypropylene. RSC Adv 4:31782–31794
29. Wang R, Lu G, Zhuang H, Yu J (2017) Synergistic catalytic effect of light rare earth element and other additives on the degree of graphitization and properties of graphite. J Mater Sci 52:663–673
30. Yuan B, Fan A, Yang M, Chen X, Hu Y, Bao C, Jiang S, Niu Y, Zhang Y, He S (2017) The effects of graphene on the flammability and fire behavior of intumescent flame retardant polypropylene composites at different flame scenarios. Polym Degrad Stabil 143:42–56
31. Yu B, Shi Y, Yuan B, Qiu S, Xing W, Hu W, Song L, Lo S, Hu Y (2015) Enhanced thermal and flame retardant properties of flame-retardant-wrapped graphene/epoxy resin nanocomposites. J Mater Chem A 3:8034–8044
32. Qiu S, Shi Y, Wang B, Zhou X, Wang J, Wang C, Gangireddy CSR, Yuen RK, Hu Y (2017) Constructing 3D polyphosphazene nanotube@mesoporous silica@bimetallic phosphide ternary nanostructures via layer-by-layer method: synthesis and applications. ACS Appl Mater Interfaces 9:23027–23038

Chapter 6
Combination of Melamine Cyanurate and Black Phosphorus for Enhancing the Flame Retardant Properties of Epoxy Resin Composites

Check for updates

6.1 Introduction

Similar to graphene and other two-dimensional (2D) layered materials, due to its excellent catalytic effect and physical barrier effect, 2D black phosphorene can be used as flame retardant nano-fillers for polymer materials [1]. Few-layer BP nanosheets can usually be obtained by ultrasonic assisted liquid phase exfoliation, mechanical exfoliation or electrochemical exfoliation [2–4]. However, due to the bare surface properties of the BP nanosheets obtained by the above stripping method, it is difficult for the BP nanosheets to be uniformly dispersed in the polymer matrix and have poor compatibility with the matrix interface. Therefore, it is difficult to achieve excellent properties of the BP/polymer composite material by directly introducing the exfoliated BP nanosheets. It is well known that the surface functionalization of 2D layered materials plays an important role in the interfacial state and performance enhancement of polymer nanocomposites [5]. Therefore, in order to solve the problem of interfacial compatibility and dispersion of BP nanosheets after their introduction into polymer matrix, functional modification is required, and it is important to maintain the geometric properties of the material while modifying it. At present, there are few reports about the preparation of 2D BP materials with excellent flame retardant effect and good dispersion. Referring to the surface modification strategy of graphene, 2D BP can be modified by covalent or non-covalent method, so as to enhance the interfacial interaction and compatibility between BP and polymer matrix, and improve the comprehensive properties of BP-based polymer nanocomposites.

Melamine cyanurate (MCA) is a kind of hydrogen bond network structure formed between cyanuric acid (CA) and melamine (MA) molecules, which can be prepared by supramolecular self-assembly method in aqueous phase [6, 7]. As a kind of conventional flame retardant, MCA is mainly composed of nitrogen element and belongs to nitrogen-based flame retardant. The flame retardant mechanism is relatively simple, the improvement of flame retardancy of polymers using MCA alone is limited. Therefore it can be combined with other types of flame retardants to improve the flame retardant efficiency of MCA [8, 9]. MCA products with different morphologies can be obtained by changing the reaction condition of synthesis process, among them, the 2D structure of MCA obtained via supramolecular self-assembly has a high specific surface area, which is similar to layered compounds, and has better flame retardant effect. However, the simple flame retardant mechanism is still a problem with the use of MCA as a flame retardant [10, 11]. Previous studies have reported that MCA/MoS_2 hybrid nanosheets forming by supramolecular self-assembly method have a significant enhancement effect on nylon 6 matrix, and a strong hydrogen bond interaction is formed between the hybrid nanosheet and the nylon 6, which obviously improves the thermal and flame-retardant properties of the nylon 6 system [12].

Generally, the combination of two or more components may have a synergistic effects, resulting in a significant improvement in the properties of polymer composites [13, 14]. In this chapter, aminated black phosphorus/melamine cyanurate (BP-NH-MCA) hybrid nanoparticles with sandwich structure were prepared by in-situ self-assembly process to achieve the above objectives. The preparation procedure as shown in Fig. 6.1. The combination of BP-NH_2 nanosheets and MCA supermolecule may show the synergistic effect of these two materials, and they play roles in different combustion stages, with different flame retardant mechanisms, which can integrate the thermal stability and flame retardant efficiency of these two components. The active group on the surface of the BP-NH-MCA hybrid material can reduce the interfacial tension between the BP-NH-MCA hybrid and the EP matrix, and improve

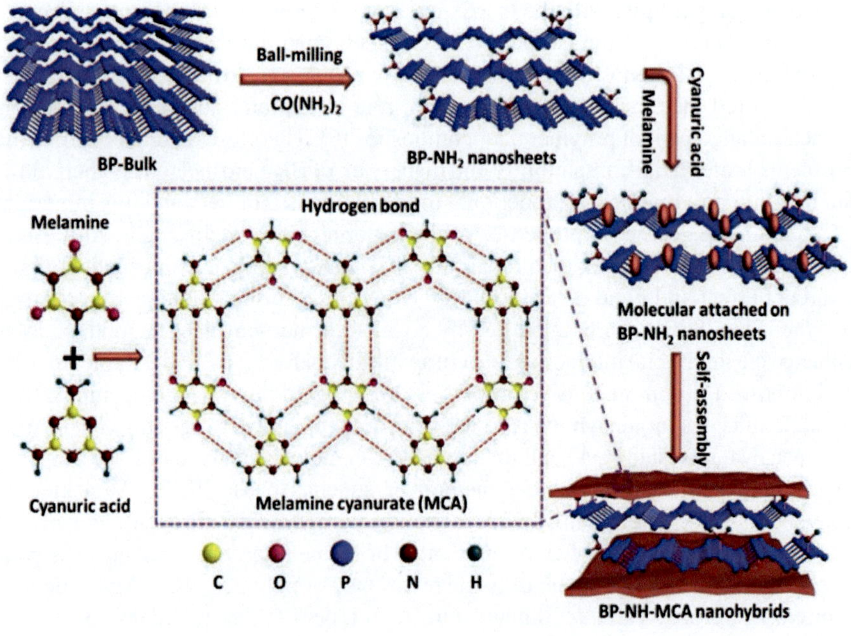

Fig. 6.1 Illustration of the fabrication process of the BP-NH-MCA

the interfacial compatibility. This chapter focuses on the effects of hybridization of $BP\text{-}NH_2$ nanosheets and MCA supramolecular on the flame retardant and smoke suppression of EP matrix.

6.2 Experimental Section

6.2.1 *Raw Materials*

Melamine ($C_3N_3(NH_2)_3$ and cyanuric acid were purchased from Aladdin Industries Ltd. (China) and used without further purification. Suppliers of other reagents have been described in Chaps. 2 and 5.

6.2.2 *Preparation of $BP\text{-}NH_2$ and BP-NH-MCA Nanohybrid*

The BP-Bulk was prepared according to the method in Chap. 2. $BP\text{-}NH_2$ was synthesized by ball milling using a simple solid-state mechanical chemistry method, and the specific experimental steps are described in Chap. 5. The BP-NH-MCA nanohybrid with sandwich structure is achieved by molecular self-assembly strategy. In order to control the amount and morphology of MCA molecules assembled on the surface of $BP\text{-}NH_2$ nanosheets, we obtained the optimal ratio of surface assembly through multiple experiments. The detailed steps are as follows: firstly, 0.4 g of $BP\text{-}NH_2$ powder was ultrasonically disperse in 1000 mL of N_2 saturated deionized water to obtain a suspension. After that, 0.29 g of melamine was added to the above $BP\text{-}NH_2$ suspension, and continuously stirred at 80 °C until the melamine was completely dissolved, and then 100 mL of an aqueous cyanuric acid solution (6.0 mg/mL) was added to the above system, the reaction was continued at 95 °C for 12 h. After the end of the reaction, the product was filtered, washed and dried overnight in a vacuum drying box. The resulted product was named BP-NH-MCA nanohybrid. Pure MCA can be synthesized in the same way, except for the step of adding $BP\text{-}NH_2$.

6.2.3 *Preparation of EP/BP-NH-MCA Nanocomposites*

The preparation procedure of EP nanocomposite with the addition of 2.0 wt% BP-NH-MCA is as follows: 0.9 g BP-NH-MCA was dispersed in 40 mL acetone with assistance of ultrasonication for 30 min. The corresponding 36.2 g of epoxy resin was then added to the above system under sonication and mechanical stirring, and uniformly mixed for 1 h. The solvent acetone was then removed from the vacuum oven at 75 °C for 4 h. Thereafter, 7.9 g of molten DDM was added to the above blend and stirred for 1 min. Then, the resin sample was cured at 100 and 150 °C for 2 h, respectively. After the curing process was completed, the EP/BP-NH-MCA2.0 sample was allowed to cool to room temperature. A similar method was used to prepare pure EP, EP/BP-NH-MCA0.5 (0.5 wt%), EP/BP-NH-MCA 1.0 (1.0 wt%) and EP/BP-$NH_2$2.0 (2.0 wt%) composites. The pure and comparative samples used in this chapter are the same as those in Chap. 5.

6.2.4 Characterization

Refer to Chaps. 2 and 3 for FTIR, XRD, Raman, XPS, TEM, SEM, TGA, TG-IR and cone calorimeter measurements.

6.3 Results and Discussion

6.3.1 Structure and Morphology of BP-NH_2 and BP-NH-MCA

The composition and structure of BP-NH_2 and BP-NH-MCA were confirmed by a series of spectra and morphology characterizations. As shown in Fig. 6.2a, the FTIR spectrum of BP-NH_2 has characteristic peaks corresponding to the intrinsic structure of the BP crystal, and these characteristic peaks can also be observed in the IR spectrum of BP-NH-MCA. After in-situ self-assembly, the IR spectrum of BP-NH-MCA is highly similar to that of pure MCA. Specifically, the strong absorption peak at 3386 cm^{-1} can be attributed to the stretching vibration of the NH_2 group on the triazine structure. The absorption peaks at 3229 and 3024 cm^{-1} are ascribed to the formation of internal hydrogen bonds of the NH_2/NH groups. The strong absorption peaks at 1738 and 1662 cm^{-1} are attributed to the shear and bending vibrations of the NH_2 group, respectively. The absorption peaks at 1533 and 1445 cm^{-1} correspond to the vibrational absorption of the C=N and C-N groups, respectively. Figure 6.2b shows the XRD patterns of MCA and BP-NH_2 as well as BP-NH-MCA nanosheets. The corresponding characteristic diffraction peaks of BP-NH_2 nanosheets correspond well with the orthorhombic crystal form of BP (PDF 65-2491). Several representative diffraction peaks at 17.1, 26.6, 34.3, 40.3, 52.5, 56.2, and 56.8° correspond to (020), (021), (040), (002), (060), (151) and (061) interlayer planes of BP, respectively. The MCA also has a corresponding crystal structure, which is consistent with the MCA standard card (JCPDS file 00-005-0217). After in-situ assembly of MCA on the surface of BP-NH_2, the spectrum of BP-NH-MCA sample contains the main diffraction peaks of MCA and BP-NH_2, and the main diffraction peak intensity of BP in BP-NH-MCA hybrid material decrease obviously. The result indicates that the MCA molecule is successfully encapsulated on the surface of the BP-NH_2 nanosheet.

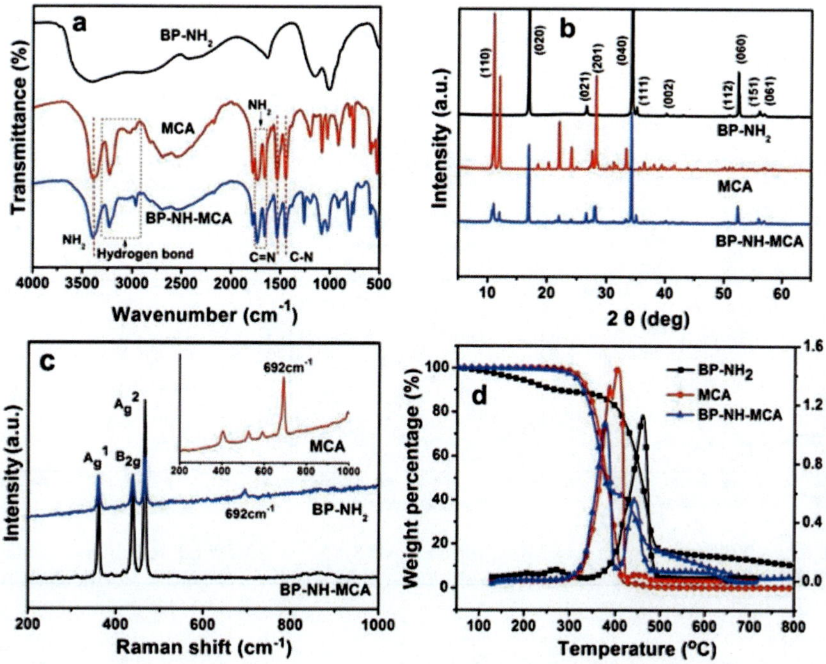

Fig. 6.2 **a** FT-IR spectra; **b** XRD patterns; **c** Raman spectra and **d** TGA curves of MCA, $BP-NH_2$ and BP-NH-MCA hybrid

Raman spectroscopy was used to analyze the structural integrity of $BP-NH_2$ and BP-NH-MCA nanosheets. As shown in Fig. 6.2c, there is a characteristic vibration peak at 692 cm^{-1} in the Raman spectrum of MCA. There are three typical vibration modes in the Raman spectrum of $BP-NH_2$, which are located at 361.2, 438.9 and 466.2 cm^{-1}, respectively, which can be attributed to the A_g^1, B_{2g} and A_g^2 vibration modes of BP. These peaks were also detected by BP-NH-MCA nanohybrid, demonstrating the integrity of the vibrational structure of $BP-NH_2$ after MCA coating. In addition, BP-NH-MCA also has vibration peaks at 692 cm^{-1}. These results further demonstrate the successful preparation of BP-NH-MCA nanohybrid. The thermal stability and structure of BP-NH-MCA can be studied by TGA under nitrogen. As shown in Fig. 6.2d, $BP-NH_2$ has a two-step mass loss process before 800 °C. The first step mass loss (<250 °C) is mainly due to the heat elimination of the $-NH_2$ groups on the $BP-NH_2$ nanosheets and the crystal defects generated during ball milling. The mass loss in second step between 400 and 550 °C is due to thermal degradation of the black phosphorus. The MCA molecule exhibited a one-step thermal degradation between 300 and 450 °C. For the BP-NH-MCA hybrid, the two-step mass loss process is mainly attributed to the thermal degradation of the MCA molecule and the thermal degradation of the black phosphorus main structure. Furthermore, the BP-NH-MCA hybrid exhibits no thermal elimination of unstable groups before 250 °C, which is

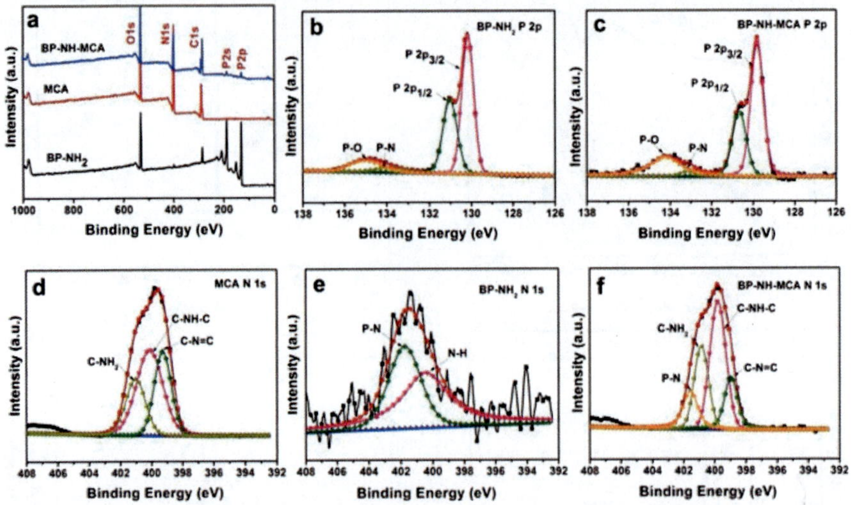

Fig. 6.3 **a** XPS survey spectra of MCA, $BP-NH_2$ and BP-NH-MCA hybrid; **b**, **c** high resolution P 2p XPS spectra of $BP-NH_2$ and BP-NH-MCA; **d–f** High resolution N 1s XPS spectra of MCA, $BP-NH_2$ and BP-NH-MCA

due to the presence of strong intramolecular and intermolecular hydrogen bonding interactions in BP-NH-MCA.

In the XPS survey spectra of $BP-NH_2$, MCA and BP-NH-MCA hybrid (Fig. 6.3a), the XPS spectrum of the $BP-NH_2$ sample shows the signals of C, O, P and N elements, and the MCA shows the signals of C, O and N elements. The signal peaks of C, O, P, and N elements can be clearly observed in the spectrum of BP-NH-MCA hybrid, and the content of N element is significantly increased compared with $BP-NH_2$. Figure 6.3b shows the P 2p peak of $BP-NH_2$, which can be deconvoluted into four peaks of 130.1, 131.0, 133.9, 134.9 eV, corresponding to the P $2p_{3/2}$ and P $2p_{1/2}$ of the P-P bond, P-N and P-O bonds, respectively. After binding to the MCA molecule, the P 2p peak of the BP-NH-MCA hybrid can also be divided into four peaks corresponding to the P-P bond (P $2p_{3/2}$ and P $2p_{1/2}$), P-N bond and P-O bond, as shown in Fig. 6.3c. In Fig. 6.3d, the high-resolution N 1s spectrum of MCA deconvolves three characteristic peaks at 399.3, 400.1, and 401.7 eV, respectively. The nitrogen with lower binding energy attributable to sp2 hybridization (C-N=C), the higher binding energy peak can be assigned to the nitrogen-based compound in the form of C-NH-C and $C-NH_2$. Compared to MCA, the N 1s spectrum of the $BP-NH_2$ sample in Fig. 6.3e can be deconvoluted into two peaks, corresponding to the P-N and N-H bonds, respectively. As shown in Fig. 6.3f, in the N 1s XPS spectrum of BP-NH-MCA, the three strong peaks at the lower binding energy can be attributed to C-N=C, C-NH-C and $C-NH_2$ bonds, respectively. In addition, the peak at the highest binding energy can be attributed to the P-N bond, and the peaks of the BP-NH-MCA hybrid slightly shifted to the lower binding energy, compared to those of pure MCA,

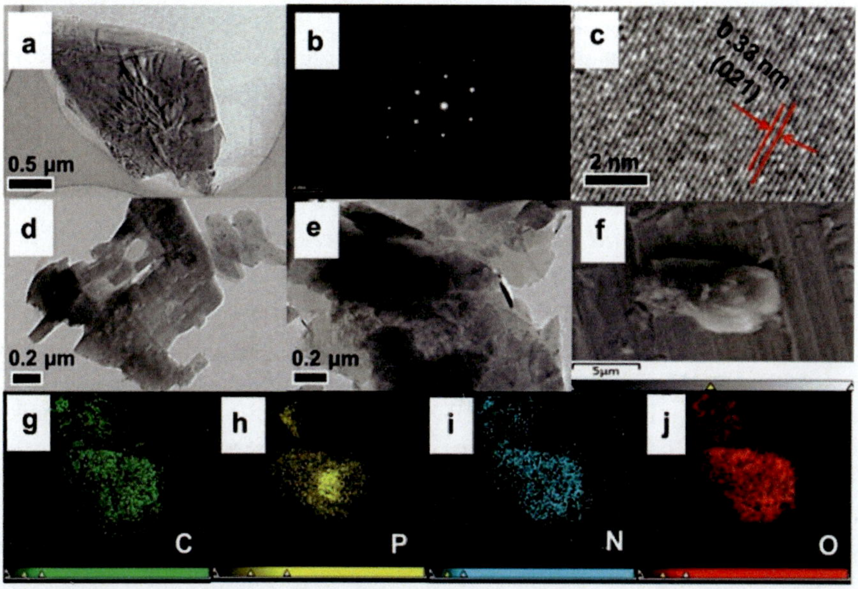

Fig. 6.4 **a** TEM image of $BP\text{-}NH_2$ nanosheets and corresponding, **b** SAED patterns and **c** HRTEM images; **d** TEM images of MCA supramolecular crystal and **e** BP-NH-MCA hybrid; **f** SEM image of BP-NH-MCA hybrid and corresponding elemental mapping images of, **g** carbon, **h** phosphorus, **i** nitrogen and **j** oxygen

this phenomenon indicates that formation of strong interaction between the MCA and the $BP\text{-}NH_2$ nanosheets.

The microstructure and elemental composition of $BP\text{-}NH_2$ and BP-NH-MCA were characterized by TEM and SEM. As shown in the TEM image (Fig. 6.4a), the $BP\text{-}NH_2$ nanosheets exfoliated by ball milling exhibit a good nanosheet morphology with a diameter of several micrometers. High-resolution TEM (HRTEM) images show that $BP\text{-}NH_2$ nanosheets maintain good crystallinity (Fig. 6.4c). Selected area electron diffraction (SAED) pattern (Fig. 6.4b) was selected to prove the orthogonal crystal structure of $BP\text{-}NH_2$ nanosheet. The 0.33 nm lattice fringe in the HRTEM can be attributed to the (021) crystal surface of BP, which is consistent with the results of XRD analysis. As shown in Fig. 6.4d, it can be seen from the TEM image of the MCA supramolecular that it exhibits a long, block-like structure with a smooth surface ranging in size from a few microns. From the TEM image of BP-NH-MCA, it can be seen that the surface of the $BP\text{-}NH_2$ nanosheet is coated with a extra phase of material, due to the MCA supramolecule attached to the surface of $BP\text{-}NH_2$ nanosheet to form a nanohybrid, which remains layered structure (Fig. 6.4e). Because the surface of $BP\text{-}NH_2$ nanosheet is rich in amino groups, through the hydrogen bond between amino groups, MA molecules are adsorbed on the surface of nanosheet and take it as a growth template to form BP-NH-MCA hybrid via the self-assembly process. In addition, Fig. 6.4f shows the SEM image of the BP-NH-MCA hybrid.

From the corresponding elemental mapping image of the SEM image, the surface of the BP-NH-MCA nanohybrid is composed of elements such as phosphorus (P), nitrogen (N), carbon (C) and oxygen (O), further validated the successful preparation of BP-NH-MCA nanohybrid, as shown in Fig. 6.4g–j.

6.3.2 *Structure and Morphology Characterization of EP/BP-NH-MCA Composites*

The interfacial compatibility between the BP-NH-MCA nanofiller and the EP matrix was investigated by cross-sectional SEM of the composite. SEM images of the frozen fracture surfaces of EP, EP/BP-$NH_2$2.0 and EP/BP-NH-MCA2.0 samples are shown in Fig. 6.5. It can be observed from the SEM image of pure EP (Fig. 6.5a) that the fracture surface of the pure epoxy is very smooth without any fracture or wrinkle morphology. As shown in Fig. 6.5b, it can be seen from the cross-sectional SEM image of EP/BP-$NH_2$2.0 that the fracture surface of the sample is relatively rough and has transverse stripes along the same direction due to strong interfacial forces. The small wrinkle morphology indicates that there is a certain interfacial

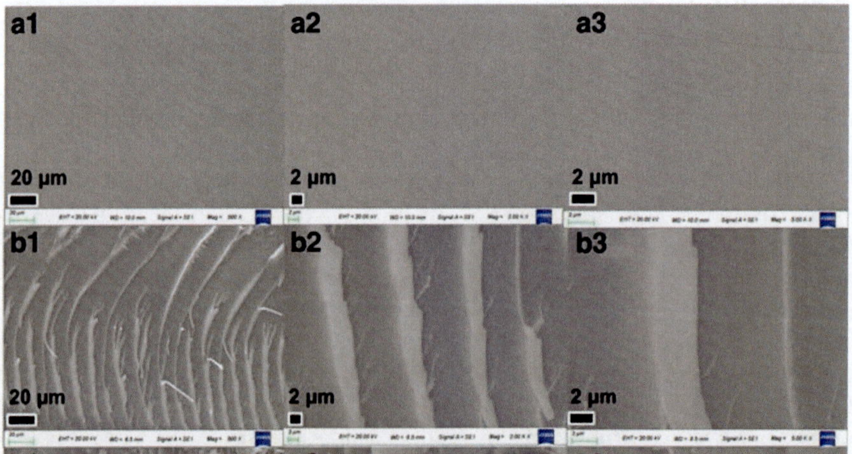

Fig. 6.5 SEM images of **a** pure EP, **b** EP/BP-$NH_2$2.0 and **c** EP/BP-NH-MCA2.0 nanocomposites at different magnifications

interaction between the BP-NH_2 nanosheet and the EP matrix. For the EP/BP-NH-MCA2.0 composite, as shown in Fig. 6.5c, compared with pure EP and EP/BP-$NH_2$2.0 composite, the surface of the sample is rougher, and the cross lines are irregular, forming a large wrinkle morphology. This may be due to the integration of strong hydrogen bonding interactions between BP-NH_2 nanosheets and MCA supramolecules resulting in enhanced interfacial interaction between BP-NH-MCA and EP matrix.

6.3.3 Thermal Property of EP/BP-NH-MCA Composites

The effect of BP-NH-MCA hybrid on the thermal stability of EP nanocomposites was investigated by TGA analysis under nitrogen (N_2) atmosphere, as shown in Fig. 6.6. The DTG curve shows that the EP/BP-NH_2 and EP/BP-NH-MCA nanocomposites exhibit the same pyrolysis behavior as pure EP and exhibit a major mass loss process

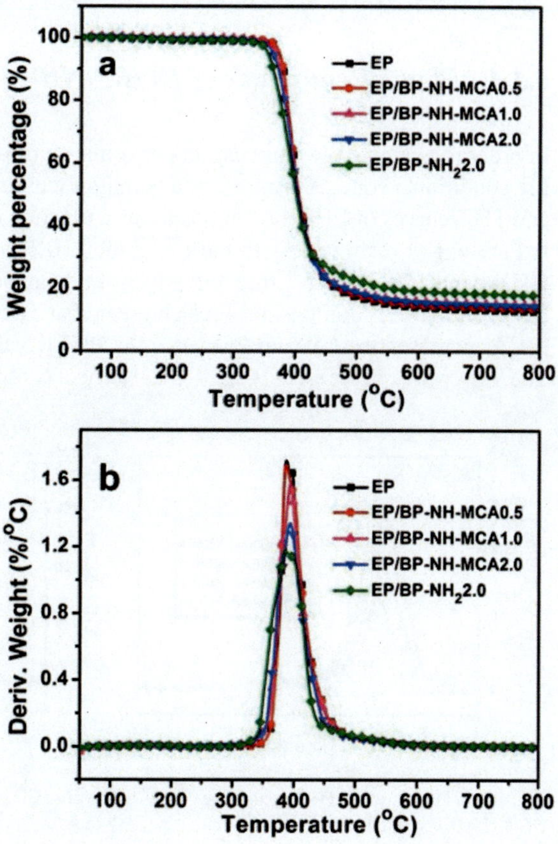

Fig. 6.6 **a** TGA curve and **b** DTG curve of pure EP and its nanocomposites

in the range of 300–500 °C. As shown in Fig. 6.6a, the initial decomposition temperature of the EP/BP-NH-MCA nanocomposite was slightly lower than that of pure EP, due to the catalytic carbonization effect of BP-NH-MCA nanosheets promote the resin matrix to degrade in advance. With the increasing content of BP-NH-MCA in epoxy resin, the char residue of nanocomposites gradually increased at 800 °C significantly, higher than that of pure EP. In particular, the char residue yield of the EP/BP-$NH_2$2.0 sample was 18.0% at 800 °C, which was slightly higher than the char residue yield of EP/BP-NH-MCA2.0 nanocomposite with the same loading. However, the initial decomposition temperature is reduced, which is caused by catalytic carbonization and physical barrier effects of BP nanosheets during the degradation process. In the DTG curve of EP nanocomposites (Fig. 6.6b), the maximum mass loss rate of EP/BP-NH_2 and EP/BP-NH-MCA composites is much lower than that of pure EP. BP-NH-MCA nanosheets act as physical barriers to inhibit the diffusion of pyrolysis products and their exchange with oxygen. BP-NH_2 nanosheets and MCA supramolecules build a phosphorus-nitrogen flame retardant system with good effect, which can promote the formation of dense carbonized char layer, thereby improving the thermal oxidative resistance of EP nanocomposites.

6.3.4 Flame Retardancy of EP/BP-NH-MCA Composites

In order to evaluate the combustion performance of polymer composites under real fire conditions, cone calorimetry is a commonly used evaluation method. The HRR and THR curves of EP nanocomposites as a function of time are shown in Fig. 6.7a, b. Pure epoxy resin is easy to burn, and the PHRR and THR values are as high as 2053.4 and 160.7 MJ/m^2, respectively. From the experimental results of a series of EP/BP-NH-MCA composites, it can be seen that as the addition amount of BP-NH-MCA increases from 0.5 to 2.0 wt%, the PHRR value decreases by 16.0–43.2%. The THR value decreased by 25.9–41.3%. As a contrast, the PHRR value and THR

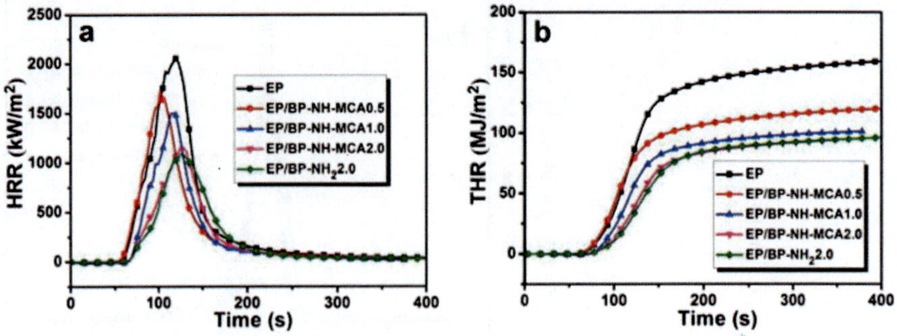

Fig. 6.7 **a** HRR and **b** THR versus time curves of the pure EP and its nanocomposites with different BP-NH-MCA contents

value decreased by 46.5% and 39.6% respectively in the combustion process due to the addition of 2 wt% BP-NH_2 in EP, which was almost the same as that of the EP/BP-NH-MCA2.0 nanocomposites with the same loading. From the point of view of the phosphorus content of the flame retardant element, the addition of BP-NH-MCA hybrid makes the phosphorus content of the composite material lower than that of the BP-NH_2 nanosheet, but can achieve considerable flame retardancy. The result demostrates that the flame retardancy of EP/BP-NH-MCA nanocomposites is significantly improved.

Figures 6.8a, b show the SPR and TSR versus time curves for EP nanocomposites. Epoxy resins have a polyaromatic ring structure that releases a large amount of flue gas during combustion, thus pure EP has high SPR and TSR values. Compared with pure EP, the SPR and TSR values of EP/BP-NH-MCA nanocomposites decrease significantly with the addition of BP-NH-MCA. EP/BP-NH-MCA2.0 sample showes lower SPR and TSR values than EP/BP-$NH_2$2.0. The addition of BP-NH-MCA hybrid significantly reduces the fire hazard of EP composites, mainly due to the physical barrier of BP-NH-MCA, in addition, the combination of BP-NH_2 and MCA supramolecules constructs a efficient synergistic phosphorus-nitrogen flame retardant system, which has synergistic catalytic carbonization effect, promoting the formation of compact continuous char layer, thus inhibiting the release of pyrolysis gas products inside the material, delaying the transfer of mass and heat in the combustion process, and protecting the unburned materials inside the polymer from direct exposure to flame during the combustion process.

The thermal degradation behavior of EP/BP-NH-MCA nanocomposites and the formation of pyrolysis products of composites were further studied by TG-FTIR technique. Figure 6.9 shows the release intensity versus time for several typical volatile pyrolysis products of EP nanocomposites. The characteristic peaks of the gaseous pyrolysis products released during the combustion of EP nanocomposites are as follow: refer to Chap. 5.

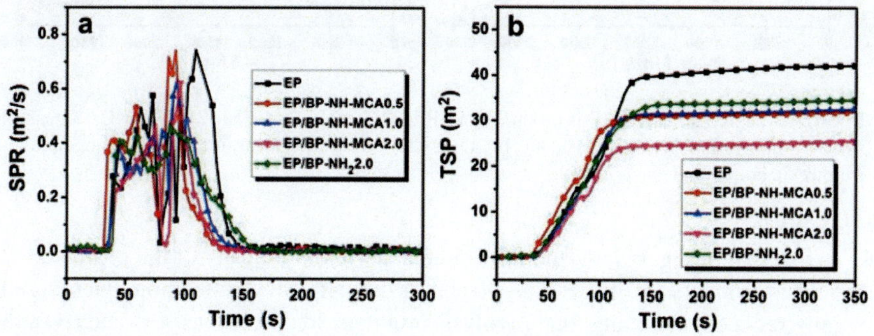

Fig. 6.8 **a** SPR and **b** TSR versus time curves of the pure EP and its nanocomposites with different BP-NH-MCA contents

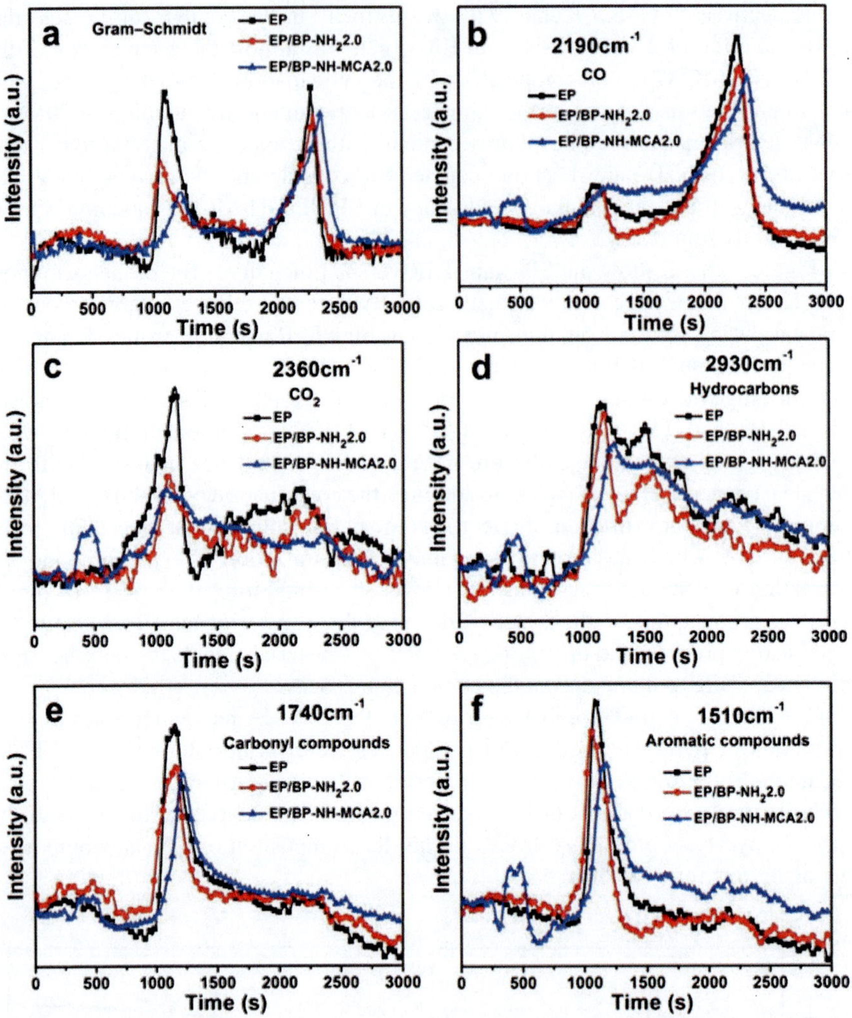

Fig. 6.9 Absorbance of pyrolysis products for EP, EP/BP-$NH_2$2.0 and EP/BP-NH-MCA2.0 sample versus time: **a** total pyrolysis product; **b** CO; **c** CO_2; **d** hydrocarbon; **e** carbonyl compound and **f** aromatic compound

As shown in Fig. 6.9a–f, the maximum absorption intensity of the characteristic pyrolysis products for EP/BP-$NH_2$2.0 and EP/BP-NH-MCA2.0 samples were significantly reduced, including total pyrolysis products, hydrocarbons, CO and CO_2, the carbonyl compound and aromatic compound, are lower than the pure EP sample. In particular, the typical volatile product of the EP/BP-NH-MCA2.0 has the lowest absorption intensity compared to the pure EP and EP/BP-$NH_2$2.0 samples. The reduction in the release of pyrolysis products, including hydrocarbons, carbonyls and aromatics, is beneficial in reducing the release of combustion smoke and heat.

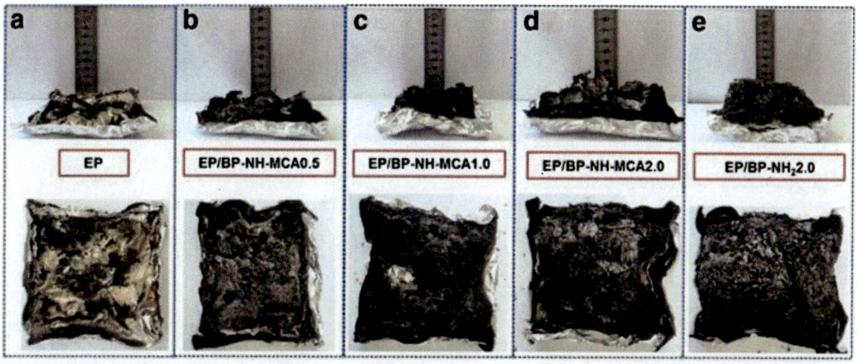

Fig. 6.10 Top view and side view digital photos of the exterior char for EP and its nanocomposites

As a major toxic gas product during combustion, the reduction in CO production also helps to reduce the toxicity of the flue gas, thereby improving the fire safety performance of the composites.

6.3.5 Condensed Phase Analysis of EP/BP-NH-MCA Composites

In order to explore the flame retardant mechanism of BP-NH-MCA hybrid, the residual char of the composite after the cone calorimeter test was studied. A top view and a side view of the char residue of EP and EP/BP-NH-MCA nanocomposite are shown in Fig. 6.10. It can be seen from Fig. 6.10a that there is a small amount of broken char fragments on the surface of the pure EP, which is caused by the loose surface of the char layer and the release of the pyrolysis gas. However, as the addition amount of BP-NH-MCA increased from 0.5 to 2.0 wt% (Fig. 6.10b–d), the volume and amount of char residue gradually increased. The digital photograph of the EP/BP-NH-MCA2.0 sample exhibited a continuous dense char layer with a tendency to swell (Fig. 6.10d). In contrast, the EP/BP-$NH_2$2.0 sample also has an expansion effect but with small pores (Fig. 6.10e).

The microstructures of the external and internal char residue of EP, EP/BP-NH_2 and EP/BP-NH-MCA nanocomposites are shown in Fig. 6.11. It can be seen from the SEM image of the outer char layer of pure EP (Fig. 6.11a) that the surface of the sample has fragmented depressions, and the deep voids can reach the interior of the char residue. By comparison, it can be found that, as shown in Fig. 6.11b, the outer char residue of the EP/BP-$NH_2$2.0 sample exhibits a tight and continuous surface, and the smooth surface has no pores. The SEM of char residue for EP/BP-NH-MCA2.0 sample showed that the surface also formed a continuous and dense char layer (Fig. 6.11c). The SEM images of the interior char residue of above samples are shown in Fig. 6.11d–f, respectively. It can be seen that the internal char residue of the

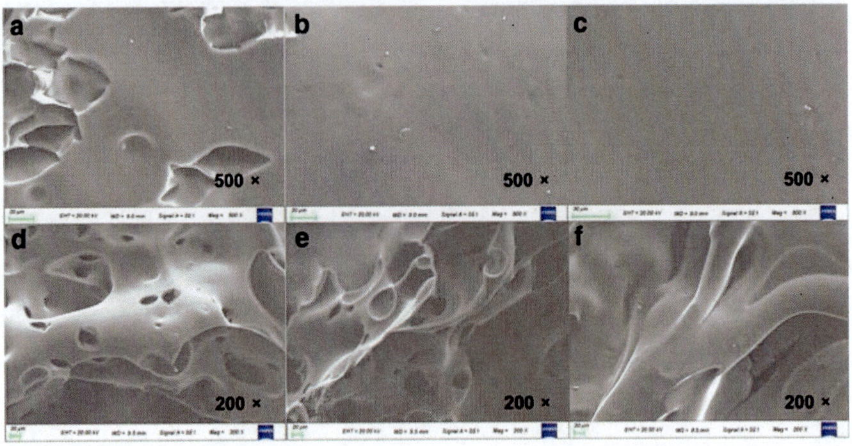

Fig. 6.11 SEM images of external **a–c** and interior, **d–f** char residues for EP, EP/BP-$NH_2$2.0 and EP/BP-NH-MCA2.0 nanocomposites

EP/BP-NH-MCA sample is more continuous and dense, compared with the internal residual char of the pure EP and EP/BP-NH_2 samples, and there is no large amount of pores. It can be observed that there are large pores form on the surface of the internal char of the pure EP and EP/BP-NH_2 samples. These phenomena indicate that the catalytic carbonization effect and physical barrier effect of BP-NH-MCA nanohybrid can contribute to the formation of high quality char layers during combustion. These continuous dense char layers can effectively delay the heat and mass transfer and the escape of pyrolysis gases, thus improve the fire safety of polymer matrix.

The composition of the char residue was analyzed by Raman spectroscopy. From Fig. 6.12, it can be observed that there are two representative characteristic peaks of 1596 and 1365 cm^{-1} in the Raman spectra of EP, EP/BP-$NH_2$2.0 and EP/BP-NH-MCA2.0 samples, defined as G peak and D peak. The area ratio of D peak to G peak (I_D/I_G) can reflect the degree of graphitization of char residue. The relatively low I_D/I_G value indicates that the high graphitization degree of char residue, which help to protect the internal materials from burning during the combustion. Specifically, the I_D/I_G value of pure EP was 3.20, the I_D/I_G value of the EP/BP-$NH_2$2.0 sample was 3.10, and the EP/BP-NH-MCA2.0 sample exhibits the lowest I_D/I_G value of 3.08. The higher graphitization degree of the char layer for this sample is due to the combination of the physical barrier and catalytic carbonization effect of the BP-NH-MCA hybrid structure in the EP combustion process, which promotes the formation of high quality char layer.

The structural composition of the char residue of EP and its nanocomposites was analyzed by XRD and XPS, as shown in Fig. 6.13. From the XRD spectra of EP and its nanocomposites (Fig. 6.13a), it can be observed that a wide diffraction peak at $2\theta = 23°$ corresponds to the (002) plane diffraction of graphite, indicating that graphitized char was formed after cone calorimetric experiment. However, no characteristic diffraction peaks of BP crystals were found in the spectra, indicating

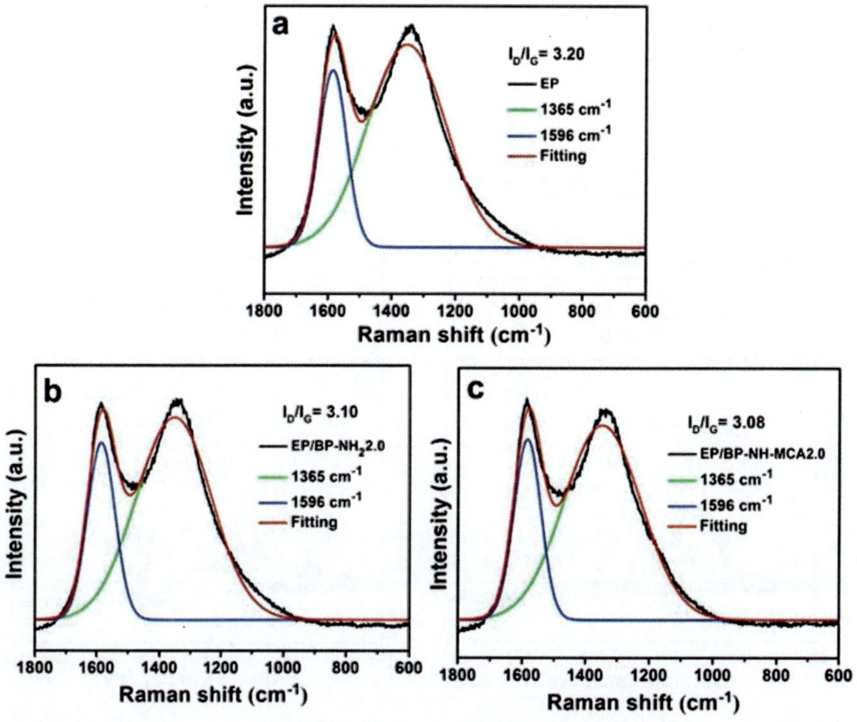

Fig. 6.12 Raman spectroscopy of char residue from EP and its nanocomposites

that the BP moiety in the BP-NH_2 and BP-NH-MCA hybrid structures completely degraded during combustion. Figure 6.13b shows the XPS measurement spectra of EP and EP/BP-NH-MCA nanocomposites. The surface of the pure EP sample consists of C, O and N elements, except for these elements, the P element appears in the XPS spectrum of the char residue of the EP/BP-NH_2 and BP-NH-MCA sample, due to the PO_x and its derivative which are produced after the decomposition of BP remain in the char residue. As shown in Fig. 6.13c, the P 2p peak of EP/BP-NH-MCA can be deconvolved into three characteristic peaks, corresponding to P-O-P, O-P=O and P_2O_5, respectively, further illustrating the decomposition of BP. Figure 6.13d shows the N 1s high resolution XPS spectrum of EP/BP-NH-MCA. The three characteristic strong peaks existing can be attributed to P-N, C-N-C and C-N=C bonds, respectively, wherein the C-N-C and C-N=C bonds belong to the structure in the MCA supramolecular, indicating that the MCA is still bonded to the char residue after being decomposed. On the other hand, the existence of P-N is due to the formation of cross-linked phosphorus oxynitride.

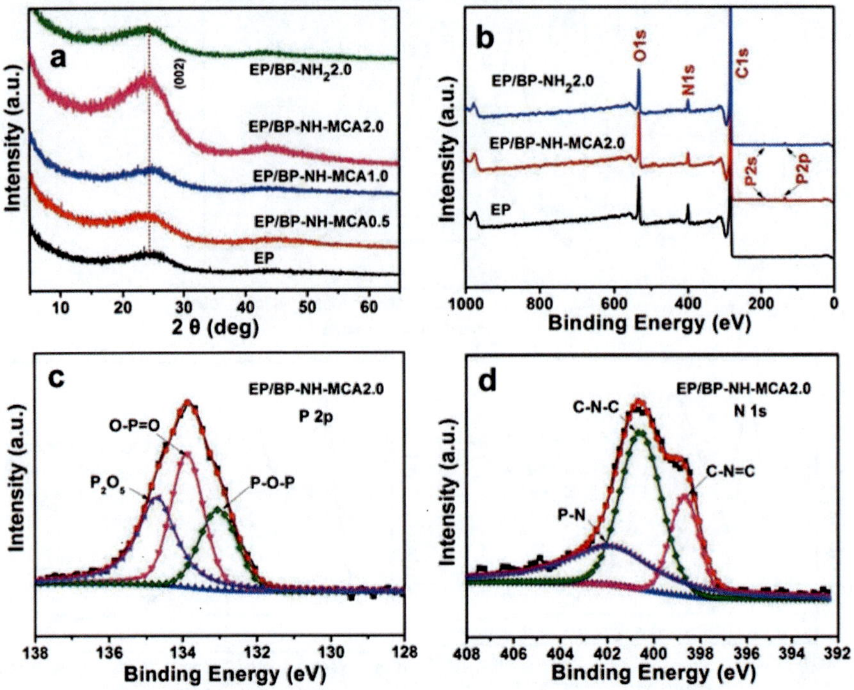

Fig. 6.13 **a** XRD spectra of char residue for EP and its composites; **b** XPS survey spectra of char residue for EP and its composites; High-resolution **c** P 2p and **d** N 1s XPS spectra of the residues for EP/BP-NH-MCA2.0 sample

6.4 Summary

In this chapter, -NH_2 functionalized BP (BP-NH_2) was synthesized by urea-assisted ball-milling. Subsequently, the MCA supramolecular/aminated black phosphorus nanohybrid (BP-NH-MCA) was prepared by in-situ self-assembly using BP-NH_2 nanosheet as a template. The structure and composition was confirmed by FT-IR, XRD, Raman and XPS. MCA supramolecules are combined with BP-NH_2 nanosheets in a non-covalent manner. Through this design, the BP-NH-MCA nanohybrid can maintain the original high specific surface area of BP and show good flame retardant effect in EP composites. Adding 2 wt % of BP-NH-MCA can obviously improve the thermal stability of EP composites. The results of cone calorimetry showed that BP-NH-MCA hybrid could effectively reduce the PHRR, THR, TSR and SPR values of the epoxy composite system. TG-FTIR analysis showed that by adding BP-NH-MCA hybrid, the volatile products including toxic CO produced from EP composites during combustion, were significantly inhibited, which was helpful to reduce the smoke toxicity. The significant improvement in flame retardancy of EP composites based on BP-NH-MCA is mainly attributed to the combination effect of catalytic carbonation and physical barrier of BP-NH_2 nanosheets and MCA supramolecules.

References

1. Qiu S, Zhou Y, Zhou X, Zhang T, Wang C, Yuen RK, Hu W, Hu Y (2019) Air-stable polyphosphazene-functionalized few-layer black phosphorene for flame retardancy of epoxy resins. Small 15:1805175
2. Kang J, Wood JD, Wells SA, Lee J-H, Liu X, Chen K-S, Hersam MC (2015) Solvent exfoliation of electronic-grade, two-dimensional black phosphorus. ACS Nano 9:3596–3604
3. Erande MB, Pawar MS, Late DJ (2016) Humidity sensing and photodetection behavior of electrochemically exfoliated atomically thin-layered black phosphorus nanosheets. ACS Appl Mater Interfaces 8:11548–11556
4. Chen Y, Jiang G, Chen S, Guo Z, Yu X, Zhao C, Zhang H, Bao Q, Wen S, Tang D (2015) Mechanically exfoliated black phosphorus as a new saturable absorber for both Q-switching and mode-locking laser operation. Opt Exp 23:12823–12833
5. Ramanathan T, Abdala A, Stankovich S, Dikin D, Herrera-Alonso M, Piner R, Adamson D, Schniepp H, Chen X, Ruoff R (2008) Functionalized graphene sheets for polymer nanocomposites. Nat Nanotechnol 3:327
6. He L, Liu Y, Lin M, Awika J, Ledoux DR, Li H, Mustapha A (2008) A new approach to measure melamine, cyanuric acid, and melamine cyanurate using surface enhanced Raman spectroscopy coupled with gold nanosubstrates. Sens Instrum Food Qual Saf 2:66–71
7. Seki T, Yagai S, Karatsu T, Kitamura A (2008) Formation of Supramolecular polymers and discrete dimers of perylene bisimide dyes based on melamine−cyanurates hydrogen-bonding interactions. J Org Chem 73:3328–3335
8. Liu Y, Wang Q (2006) Melamine cyanurate-microencapsulated red phosphorus flame retardant unreinforced and glass fiber reinforced polyamide 66. Polym Degrad Stabil 91:3103–3109
9. Gijsman P, Steenbakkers R, Fürst C, Kersjes J (2002) Differences in the flame retardant mechanism of melamine cyanurate in polyamide 6 and polyamide 66. Polym Degrad Stabil 78:219–224
10. Yang W, Song L, Hu Y, Lu H, Yuen RK (2011) Enhancement of fire retardancy performance of glass-fibre reinforced poly (ethylene terephthalate) composites with the incorporation of aluminum hypophosphite and melamine cyanurate. Compos Part B Eng 42:1057–1065
11. Yanagimoto A, Kumazawa S, Takakuwa Y (1979) Melamine cyanurate as a flame retardant agent. In: Google Patents
12. Feng X, Wang X, Cai W, Hong N, Hu Y, Liew KM (2016) Integrated effect of supramolecular self-assembled sandwich-like melamine cyanurate/MoS_2 hybrid sheets on reducing fire hazards of polyamide 6 composites. J Hazard Mater 320:252–264
13. Peeterbroeck S, Alexandre M, Nagy J, Pirlot C, Fonseca A, Moreau N, Philippin G, Delhalle J, Mekhalif Z, Sporken R (2004) Polymer-layered silicate-carbon nanotube nanocomposites: unique nanofiller synergistic effect. Compos Sci Technol 64:2317–2323
14. Zhu C, Zhai J, Wen D, Dong S (2012) Graphene oxide/polypyrrole nanocomposites: one-step electrochemical doping, coating and synergistic effect for energy storage. J Mater Chem 22:6300–6306

Chapter 7
Conclusions and Recommendations

7.1 Conclusions

In this thesis, the recent research progress of polymer/black phosphorene composites was reviewed. On this basis, it can be found that, at present, there are few reports on black phosphorene-based polymer composites. What's more, there is no systematic study on the effective preparation of two-dimensional black phosphorene suitable for reinforcing polymer materials. Moreover, the pure two-dimensional black phosphorene is difficult to disperse in the polymer matrix, and generally has poor interfacial compatibility with the polymer material, which limits its mechanical reinforcement and flame retardant effect. Next, the black phosphorene itself is unstable in the air, which makes the research development of the BP-based polymer composites is limited. Finally, if black phosphorus is used as a flame retardant, the current production is too small to meet the needs of practical applications. In view of the above problems, we employed several different methods to exfoliate two-dimensional black phosphorene. At the same time, the surface functionalization of black phosphorene was realized. The surface-coating modification of liquid-phase-exfoliated black phosphorus was carried out by using organic-inorganic polyphosphazene material. It was found that the functionalized black phosphorene could be uniformly dispersed in the epoxy resin matrix to improve the compatibility with the matrix. At the same time, it also inhibited the surface oxidation of black phosphorene. The exfoliation and functionalization of black phosphorene were realized by phytic acid-assisted electrochemical method, through this method, the two-dimensional black phosphorene could be produced on a large scale for preparing the polymer-based nanocomposite. The small molecule assisted ball milling method was used to realize the simultaneous exfoliation and functionalization of the two-dimensional black phosphorene. The black phosphorene having reactive groups was suitable for further surface modification, and improved its interfacial compatibility with the matrix by hydrogen bonding or covalent bonding. In addition, two-dimensional black phosphorene nanosheets were used as synthetic templates and synergistic flame retardants to prepare black phosphorene-based nano-hybrid flame retardants. These hybrid flame retardants not only showed good matrix dispersion, moreover, the thermal stability, flame retardancy and smoke suppression performance of the polymer matrix could be significantly enhanced. The research work mainly included the following four parts.

1. A two-dimensional black phosphorene was obtained from bulk black phosphorus by liquid-phase-exfoliation, On the basis of a design concept of in-situ formation of a protective layer and surface-modification of black phosphorene, we developed a cross-linked polyphosphazene functionalized BP with abundant -NH_2 groups via in-suit polycondensation of 4,4′-diaminodiphenyl ether (ODA) and hexachlorocyclotriphosphazene (HCCP) on the surface of BP nanosheets. The functionalized black phosphorene was successfully dispersed into the epoxy resin matrix by solution blending and heat curing. Due to the surface functionalization, the 2D black phosphorene in epoxy composites presented a completely delaminated and uniform dispersion state. The thermal stability, mechanical properties and flame retardant properties of polymer composites with functionalized black phosphorene have been significantly enhanced. At different stages of combustion, black phosphorene has different flame retardant mechanisms. Before the pyrolysis of black phosphorene begins, the well-dispersed two-dimensional black phosphorene could effectively inhibit the release of pyrolysis products of the epoxy resin during the pyrolysis process, and significantly reduced the PHRR and THR values of the composite. After the pyrolysis of black phosphorene, the capture of free radicals and the char-forming effect could promote the formation of a dense and thermally insulating char layer, thereby reducing the fire hazard of nanocomposites. Meanwhile, the EP/BP-PZN nanocomposites exhibited air stability after exposure to ambient conditions for four month. The air stability of the BP nanosheets in EP matrix was assigned to surface wrapping by polyphosphazene and embedding in the polymer matrix as dual protection.
2. The crucial step towards practical application of BP is the scalable preparation of single- layer or few-layer BP nanosheets. We utilized a facile, green and scalable electrochemical strategy for generating cobaltous phytate functionalized BP nanosheets (BP-EC-Exf). By means of electrochemical method, the exfoliation and functionalization of black phosphorene were realized simultaneously. Through structural characterization, it could be found that cobalt phytate was attached to the surface of the black phosphorene nanosheets by electrostatic adsorption of non-covalent bonds. Using the as-modified black phosphorene, a high performance polyurethane/BP-EC-Exf (PUA/BP-EC-Exf) nanocomposite was prepared by a convenient UV curing method for the first time. The addition of black phosphorene could significantly enhance the thermal stability, flame retardancy and mechanical properties of composites. The two-dimensional morphology, intrinsic high strength and good interfacial compatibility of the two-dimensional black phosphorene were effective guarantees for the enhancement effect. Similar to the reinforcement effect of graphene,

the significantly enhanced mechanical properties of black phosphorene-based polyurethane composite membrane could prove that the two-dimensional black phosphorene has potential application value in reinforcing polymer materials.

3. Natural nacre offers an optimized guideline for assembling two dimensional (2D) nanosheets into high performance nanocomposites with light weight, high strength and excellent mechanical properties. Considering the design principles for green and sustainable use, inspired by the "brick-and-mortar" layered structure of nacre, a multifunctional bioinspired nanocomposites of few layer hydroxyl functionalized black phosphorus (BP-OH) with nanofibrillar cellulose (NFC) were fabricated via a vacuum-assisted filtration self-assembly procedure. As a result of hydrogen bond enhancement interface interaction between 2D BP-OH and one-dimensional (1D) NFC, the effective synergistic strengthening effect of the novel nacre-like BP-OHx/NFC composite film has been successfully achieved. The addition of hydroxylated black phosphorene can significantly improve the tensile strength and elongation at break of NFC composite films. Moreover, these nacre-like composite films showed high thermal stability and excellent heat insulation and fire resistance. The nacre-inspired approach demonstrates a promising strategy for the design of the high performance and flexible BP-OHx/NFC composite film for the first time; these bioinspired materials have promising potential in practical applications, such as flexible construction materials and flame retardant insulation materials.
4. In order to overcome the single flame retardant mechanism of black phosphorene and verify the synergistic effect of two-dimensional black phosphorene and nitrogen containing flame retardants, organic-inorganic hybrid nano-flame retardants of triazine based organic framework/black phosphorene were in situ synthesized based on two-dimensional black phosphorene template to improve the flame retardant performance of epoxy resin. $-NH_2$ modified black phosphorene can provide active sites for the in situ growth of organic framework on the surface of two-dimensional black phosphorene. The abundant amino groups on the surface of the nano-hybrid flame retardant can promote its uniform dispersion in the epoxy resin. Compared with the addition of black phosphorene alone, EP/BP-NH-TOF nanocomposites have better thermal stability, flame retardant performance and smoke suppression and toxicity reduction can be attributed to the following aspects: The formation of covalent bonds improves interfacial interactions, as well as the integrated flame retardant mechanism (inherent flame retardant effect of black phosphorene, physical barrier effect of black phosphorene nanosheet and TOF nanosheet). In addition, by analyzing the char residue of condensed phase, it can be found that the synergistic carbonization effect of black phosphorene and TOF nanosheet can form a dense and expanded adiabatic char layer on the surface of the polymer material, which can improve its fire safety performance.
5. Based on the design principle of phosphorus-nitrogen synergistic effect, the flame retardant performance of polymer composites was improved by the hybrid combination of 2D black phosphorene and traditional flame retardant MCA. Based on the template of BP-NH_2 nanosheet, MCA supramolecular/BP-NH_2

(BP-NH-MCA) nanohybrid was prepared by in situ self-assembly method. The BP-NH_2 nanosheet has abundant amino group, which is conducive to the self-assembly of MCA supramolecular on its surface, and can be effectively combined by hydrogen bond interaction. This kind of inorganic-organic hybrid flame retardant can maintain the high specific surface area of the original BP with 2D morphology. The addition of BP-NH-MCA can significantly improve the thermal stability and flame retardant performance of epoxy resin. The property enhancement of the composite is mainly attributed to interface interaction of hydrogen bond and non-covalent bond, and the inherent flame retardant effect and physical barrier effect of MCA and BP-NH_2 nanosheet.

7.2 Contributions

1. The non-covalent method was used to achieve the surface modification of black phosphorene, which proved that the surface coating of polyphosphazene could improve the dispersion of two-dimensional black phosphorene in the polymer. The epoxy nanocomposites based on functionalized black phosphorene were prepared for the first time, the flame retardant mechanism of black phosphorene in the combustion process of epoxy resin composites was proposed. At the same time, it has proved that the surface wrapped black phosphorene has higher air stability in the polymer matrix.
2. Based on the application requirements of the preparation of two-dimensional black phosphorene/polymer nanocomposites, a green, scalable and improved electrochemical method was adopted to achieve the stripping and surface functionalization of black phosphorus simultaneously. Considering that the structure of black phosphorene cannot be destroyed during processing, the polymer nanocomposites were prepared by UV curing method. According to the enhancement effect of graphene and other two-dimensional materials, the polymer composites based on two-dimensional black phosphorene have obvious enhancement effect, which indicates that two-dimensional black phosphorene has potential research and application value in the field of reinforcing polymer composites.
3. Based on the layered structure of two-dimensional black phosphorene and the design principle of green and biological basis, a nacre type composite film composed of nanofiber cellulose (NFC) and few-layer hydroxyl functionalized black phosphorus (BP-OH) was prepared. The interaction between 2D BP-OH and 1D NFC is enhanced by hydrogen bonding, and the composite film exhibits an effective synergistic enhancement effect. Compared with similar nacre structures, the new composite film based on black phosphorene has excellent mechanical properties, as well as excellent thermal insulation and fire resistance. High performance flexible BP-OHx/NFC composite membrane based on black phosphorene provides a new idea for the field of biomimetic materials.

4. Considering the phosphorus-nitrogen synergistic flame retardant effect, triazine based organic framework/black phosphorene nanohybrid materials were in situ synthesized by using 2D black phosphorene as template. The study on the pyrolysis products of gas phase and the char layer of condensed phase proves the synergistic effect of two-dimensional black phosphorene and nitrogenous organic framework in improving the flame retardant performance of the polymer. This hybrid material not only has significant physical barrier effect in the combustion process of polymer, but also plays a catalytic role in the formation of char layer.
5. In order to overcome the single flame retardant mechanism, 2D black phosphorene was combined with traditional flame retardant MCA, and 2D black phosphorene was used as a template to in situ synthesize MCA supramolecular/black phosphorene nanohybrid. The influencing factors of the performance enhancement for the composite are mainly attributed to the interface interaction of hydrogen bond and non-covalent bond, and the inherent flame retardant effect and physical barrier effect of MCA and BP-NH_2 nanosheet.

7.3 Shortcomings

In this thesis, the exfoliation, surface functionalization and the preparation of polymer nanocomposites based on black phosphorene are studied, and the effects on thermal, mechanical and combustion properties of the polymer materials are studied.

The methods of surface functionalization, the intrinsic relationship between the interfacial interaction of polymer composites and the enhancement of properties are studied. The mechanism of improving the fire safety of 2D black phosphorene was discussed from two aspects of gas phase pyrolysis products and condensed phase char residue. Some progress has been made in the current research work, but due to limited time, there are still some deficiencies:

1. Due to the limited experimental conditions and time, the method to study the flame retardant mechanism of 2D black phosphorene in the polymer system is relatively simple.
2. In the fire risk assessment of polymer system, studies on the smoke toxicity are limited, and the influence mechanism of 2D black phosphorene on the smoke generation of composite materials has not been clarified.
3. Due to the consideration of the air stability of 2D black phosphorene, the type of selected polymer matrix is limited, which requires attention to other polymer systems besides those studied in this thesis.

7.4 Recommendations for Further Research Studies

1. In order to realize the practical application of black phosphorus, a low-cost, environment-friendly and large-scale method for the preparation of high-quality 2D black phosphorene is considered, and the thickness and size of 2D black phosphorene can be controlled.
2. The flame retardant mechanism of 2D black phosphorene in the combustion process of its polymer composites was further studied by some advanced means. Such as pyrolysis gas chromatography, electron spin resonance, etc., from the molecular or atomic level to clarify the flame retardant mechanism.
3. Strengthen the research on the toxicity of flue gas, use the steady-state tubular furnace and other equipment to evaluate the fire risk of composite materials, and study the variation law of CO concentration and smoke density generated by combustion.
4. Study the influence of 2D black phosphorene on the combustion performance of other polymer systems, so as to more systematically study the combustion performance of polymer composites based on 2D black phosphorene.

About the Author

Dr. Shuilai Qiu obtained his Ph.D. Degree at University of Science and Technology of China, Hefei, China in 2019. He is now in State Key Laboratory of Fire Science, University of Science and Technology of China. Dr. Qiu has published 17 journal articles and 3 conference articles as the first and corresponding author, and holds 2 patents on the flame retardant materials. He has been granted Excellent Doctoral Dissertation Award of Chinese Academy of Sciences (2020) and President of the Chinese Academy of Sciences Special Award (2019).